Lecture Notes in Mathematics

Edited by A. Dold and B. Eckmann

954

Sudhakar G. Pandit
Sadashiv G. Deo

Differential Systems Involving Impulses

Springer-Verlag
Berlin Heidelberg New York 1982

Authors
Sudhakar G. Pandit
Sadashiv G. Deo
Department of Mathematics
Centre of Post-graduate Instruction and Research
University of Bombay
Panaji, Goa – 403 001, India

AMS Subject Classifications (1980): 34 A XX, 34 C 11, 34 D XX, 34 H 05, 45 F XX, 93 D XX

ISBN 3-540-11606-0 Springer-Verlag Berlin Heidelberg New York
ISBN 0-387-11606-0 Springer-Verlag New York Heidelberg Berlin

Printing and binding: Beltz Offsetdruck, Hemsbach/Bergstr.
2146/3140-543210

PREFACE

When a system described by an ordinary differential equation is subjected to perturbations, the perturbed system is again an ordinary differential equation in which the perturbation function is assumed to be continuous or integrable, and as such, the state of the system changes continuously with respect to time. However, in many physical problems (optimal control theory in particular), one can not expect perturbations to be well behaved. Biological systems such as heart beats, blood flows, pulse frequency modulated systems and models for biological neural nets exhibit an impulsive behaviour. Therefore, perturbations of impulsive type are more realistic. This gives rise to Measure Differential Equations. The derivative involved in these equations is the distributional derivative. The fact that their solutions are discontinuous (they are functions of bounded variation), renders most of the classical methods ineffective, thereby making their study interesting.

The systems involving impulsive behaviour are in abundance. We mention below some problems of this kind.

(i) Growth Problem : A fish breeding pond maintained scientifically is an example of this kind. Here the natural growth of fish population is disturbed by making catches at certain time intervals and by adding fresh breed. The natural growth of fish population is disturbed at some time intervals. This problem therefore involves impulses. We study such a model in some details in Chapter 1.

(ii) Case and Blaquiere Problem [2, 4] : The profit of a roadside inn on some prescribed interval of time $\tau \leq t \leq T$ is a function of the number of strangers who pass by on the road each day and of the number of times the inn is repainted during that period. The ability to attract new customers into the inn depends on its appearance which is supposed to be indexed by a number x_1. During time intervals between paint jobs, x_1 decays according to the law

$$x_1' = -k\,x_1 \,, \quad k = \text{positive constant.}$$

The total profit in the planning period $\tau \leq t \leq T$ is supposed to be

$$W(T) = A \int_{\tau}^{T} x_1(t)dt - \sum_{\alpha=1}^{N(T)} C_\alpha$$

where $N(T)$ is the number of times the inn is repainted, C_α, $\alpha = 1,\ldots, N(T)$, the cost of each paint job and $A > 0$ is a constant. The owner of the inn wishes to maximize his total profit or equivalently to minimize $- W(T)$. The problem has been treated in several details in [2,4].

(iii) <u>Control Problem</u> : In an optimal control problem given by a system

$$x' = f(t, x, u)$$

representing certain physical process, the central problem is to select the function $u(t)$ from a given set of controls so that the solution $x(t)$ of the system has a preassigned behaviour on a given time interval $[t_0,T]$ so as to minimize some cost functional. Suppose that the control function $u(t)$ has to be selected from the set of functions of bounded variation defined on $[t_0,T]$, then the solution $x(t)$ of the control system may possess discontinuities. Hence the given control problem has to be represented by a differential equation involving impulses.

(iv) <u>Ito's Equation</u> [15,50] : The dynamic system of control theory is representable by ordinary vector differential equation of the form

$$\frac{dx}{dt} = f(t, x), \quad t_0 \leq t \leq T$$

where $x \in R^n$ and f is such that the system admits a unique solution $x(t, x_0)$, $t_0 \leq t \leq T$ for the initial state $x(t_0) = x_0 \in R^n$.

The Ito's stochastic differential equation is of the form

$$dx = f(t, x)dt + G(t, x)\, dw(t), \quad t_0 \leq t \leq T$$

where f is chosen as above, $G(t, x)$ is an $n \times m$ matrix valued function of (t, x), W is a separable Wiener process (Brownian motion) in Euclidean m-space. This equation is equivalent to a stochastic integral equation

$$x(t) = x(t_0) + \int_{t_0}^{t} f(s, x(s))ds + \int_{t_0}^{t} G(s,x(s))dw(s)$$

defined on $I = [t_0,T]$, $(T < \infty)$. Assume that f and G are measurable in (t,x) for $t \in I$ and $x \in R^n$ and satisfy certain growth conditions. It is to be noted that

$$\int_{t_0}^{t} G(s, x(s))dw(s) = \lim_{\gamma \to \infty} \sum_{j=0}^{\gamma - 1} G(t_j, x(t_j))[w(t_{j+1})-w(t_j)]$$

where $t_0 \leq t_1 \leq \ldots \leq t_\gamma = t$ and that t_j become dense in $[t_0,t]$ as $\gamma \to \infty$.

Ito's equation has been studied in several details in [15,50]. The stochastic integral equation is closely related to the measure integral equation studied in Chapter 2 subsequently.

In the classical analysis of the differential systems, solutions are generally continuous functions while in the case of impulsive systems, solutions are functions of bounded variations. Hence the methods of classical analysis are not sufficient to describe the impulsive behaviour of systems. It is therefore necessary to study existence, uniqueness and continuation of solution, stability, boundedness criteria in the case of linear and nonlinear differential systems involving impulses.

In this monograph an attempt is made to unify the results from several research papers published during the last fifteen years. We feel that a monograph of this kind would assist the readers to reach to the mainstream of this area of research for making new contributions to the subject and employing the techniques in physical models.

The monograph is divided into five chapters and deals with the problems of existence, uniqueness, stability, boundedness and asymptotic equivalence associated with measure differential equations. Chapter 1 is on preliminaries. Chapter 2 contains results on existence and uniqueness. A control problem is also considered. In Chapter 3, fixed-point theorems and generalized integral inequalities are employed to derive results on stability and asymptotic equivalence. To study the predominant effect of the impulses, it is necessary to consider systems wherein impulses occur not only in perturbation terms but are also present in the original system itself. Chapter 4 provides this study. Chapter 5 deals with the extension of Lyapunov's second method to the study of impulsive systems.

The notes given at the end of each chapter indicate the sources which have been consulted. Some research papers which are closely related but not included are given for guidance and completeness. The q.e.d. mark is used to indicate the end of the proof of a result. Complete bibliography is given at the end.

We are grateful to the Department of Atomic Energy, Bombay, India for providing the financial support to this work. Our thanks are due to the referee of this monograph for fruitful comments and to Mr. M. Parameswaran of Indian Institute of Technology, Bombay, for quick and efficient typing of the manuscript.

Panaji, Goa
April, 1982

S. G. PANDIT
S. G. DEO

C O N T E N T S

CHAPTER 1

PRELIMINARIES

The purpose of this chapter is to make the reader familiar with the prerequisites needed for the study of measure differential equations.

1.1 The Space BV(J).

Let $J = [\, t_o, \infty\,)$, $t_o \geq 0$ and R^n denote the Euclidean n-space. The norm of $x = (\, x_1, x_2, \ldots, x_n) \in R^n$ is defined as

$$|x| = \sum_{i=1}^{n} |x_i| ,$$

whereas that of an n by n matrix $M = (m_{ij})$ as

$$|M| = \sum_{i=1}^{n} \sum_{j=1}^{n} |m_{ij}| .$$

Let f be a function defined on the set of real numbers and taking values in R^n. Consider all possible partitions $\pi : a = t_o < t_1 < \ldots < t_N = b$ of an interval $[a,b]$ in R. The quantity

$$V(f,[a,b]) = \sup_{\pi} \{ \sum_{i=1}^{N} |f(t_i) - f(t_{i-1})| \}$$

where π runs over the set of all partitions of $[a,b]$, is defined as the total variation of f on $[a,b]$. f is said to be of bounded variation on $[a, \infty)$ if f has bounded variation on each interval $[a,t]$, $a \leq t < \infty$ and the set $V(f,[a,t])$ of total variations is bounded. In this case

$$V(f,[a,\infty)\,) = \sup_{t \geq a} V(f,[a,t]).$$

The space of all functions of bounded variation on J and taking values in R^n is denoted by $BV(J) = BV(J,R^n)$. The norm of $f \in BV(J)$ is defined by

$$\| f \|^* = V(f,J) + |f(t_o+)|.$$

Under this norm, $BV(J)$ is a Banach space.

In the space $BC(J)$ of bounded, continuous functions on J, compact subsets are characterized by the well-known Ascoli-Arzela theorem. A similar, but rather weak result in $BV(J)$ is the following :

<u>Theorem 1.1.</u> Let $\mathfrak{F}$ be an infinite family of functions $\in BV([a,b])$ such that all functions of the family together with their total variations are uniformly bounded. Then there exists a sequence $\{\phi_k\}$ in $\mathfrak{F}$ which converges at every point of $[a,b]$ to some function $\phi \in BV([a,b])$; moreover,

$$V(\phi,[a,b]) \leq \liminf_{k \to \infty} V(\phi_k,[a,b]).$$

An important property of a function of bounded variation is that it is differentiable almost everywhere. Moreover, the set of its points of discontinuity is at most countable. Quite often, we shall be interested in the following type of functions of bounded variation :

<u>Definition 1.1.</u> A function $u : J \to R$ is said to be of type $\mathcal{G}$ if

(i) u is a right-continuous function which is of bounded variation on every compact subinterval of J;

(ii) the discontinuities $t_1 < t_2 < \dots$ are isolated and are such that $t_1 > t_o$ and $t_k \to \infty$ as $k \to \infty$;

(iii) u is differentiable on each subinterval $[t_k, t_{k+1})$, $k = 1,2,\dots$ of J, where the derivative at t_k is to be understood as the right hand derivative.

One of the useful properties of the function of bounded variation is that if $[a,b]$ is a closed interval in R and f be a function of finite variation on $[a,b]$ then $f = g + h$ where g is absolutely continuous on $[a,b]$ and $h' = 0$ a. e on $[a,b]$. Further this decomposition of f is unique except for additive constants.

1.2 Complex Measures.

Let Σ be a σ - algebra of subsets of a set X. A countable collection $\{E_i\}$ in Σ such that

$$E = \bigcup_{i=1}^{\infty} E_i \ , \quad E_i \cap E_j = \phi$$

for $i \neq j$, is called a partition of E. A complex-valued set function μ on Σ such that

$$\mu(E) = \sum_{i=1}^{\infty} \mu(E_i), \qquad E \in \Sigma$$

for every partition $\{E_i\}$ of E is called a complex measure on Σ. Real measures (usually called as signed measures) are defined similarly.

The total variation measure $|\mu|$ of a complex measure μ on Σ is a set function on Σ defined by

$$|\mu|(E) = \sup \{ \sum_{i=1}^{\infty} |\mu(E_i)| \ , E \in \Sigma \}$$

where the supremum runs over all partitions $\{E_i\}$ of E. It follows that $|\mu|$ is a positive measure (usually referred to as measure) on Σ, with the property that $|\mu|(X) < \infty$. If μ is a positive measure, then of course $|\mu| = \mu$.

Let μ be a real measure on Σ. Define

$$\mu^+ = (|\mu| + \mu)/2, \quad \mu^- = (|\mu| - \mu)/2.$$

μ^+ and μ^- are respectively called positive and negative variation measures of μ, and are positive measures on Σ. Further,

$$\mu = \mu^+ - \mu^- , \ |\mu| = \mu^+ + \mu^-.$$

The integral of a function f with respect to μ is defined as

$$\int_E f \, d\mu = \int_E f d\mu^+ - \int_E f d\mu^-.$$

f is said to be μ-integrable if it is both μ^+ - and μ^- - integrable.

Let μ be a complex measure on Σ. Then there is a measurable function g such that $|g(x)| = 1$ for all $x \in X$ and

$$\mu(E) = \int_E g d|\mu|, \quad E \in \Sigma.$$

Therefore, integration with respect to a complex measure μ may be defined by the formula

$$\int_E f d\mu = \int_E f g d|\mu|, \quad E \in \Sigma.$$

Definition 1.2. Let λ be a positive measure and μ any arbitrary (positive or complex) measure on a σ-algebra Σ. μ is said to be absolutely continuous with respect to λ if, for every $\varepsilon > 0$, there exists a $\delta > 0$ such that $|\mu(E)| < \varepsilon$ for all $E \in \Sigma$ with $\lambda(E) < \delta$.

If X is a topological space, there exists a smallest σ-algebra $\mathcal{B}$ in X such that every open set in X belongs to $\mathcal{B}$. Members of $\mathcal{B}$ are called Borel sets of X. A measure μ defined on the σ-algebra of all Borel sets in a locally compact Hausdorff space X is called a Borel measure on X.

Definition 1.3. A complex Borel measure μ defined on the σ-algebra $\mathcal{B}$ of all Borel sets in a locally compact Hausdorff space X is said to be regular if for each $E \in \mathcal{B}$ and $\varepsilon > 0$, there exist a set $F \in \mathcal{B}$ whose closure is contained in E and a set $G \in \mathcal{B}$ whose interior contains E such that $|\mu(C)| < \varepsilon$ for every $C \in \mathcal{B}$ with $C \subset G - F$.

Let f be a right continuous function on an open interval $I = (a,b)$. Extend the domain of f to $[a,b]$ by defining $f(a)=f(b)= 0$. The set function μ defined by $\mu([a,d]) = f(d) - f(a)$ and $\mu((c,d]) = f(d) - f(c)$, for $a < c < d \leq b$ has a regular, countably additive extension to the σ-algebra of all Borel sets in $[a,b]$. The restriction of this extension to the σ-algebra of Borel subsets of I is called the Borel - Stieltjes measure in I determined by f. Now let $\mathcal{B}^*$ consist of all sets of the form $E \cup N$ where $E \in \mathcal{B}$ and N is a subset of a set $M \in \mathcal{B}$ with $|\mu|(M) = 0$. Then $\mathcal{B}^*$ is a σ-algebra and if the domain of μ is extended to $\mathcal{B}^*$ by defining $\mu(E \cup N) = \mu(E)$, the extended function is countably additive on $\mathcal{B}^*$.

The function μ with domain $\mathcal{B}^*$ is called the Lebesgue-Stieltjes measure determined by f, and the integral $\int_I g\, d\mu$ is written as

$$\int_a^b g(t)df(t).$$

Since $|\mu|(I') = V(f,I')$ if I' is any interval in I, the integral $\int_I g\, d|\mu|$ is written as

$$\int_a^b g(t)dVf(t),$$

where Vf is the total variation function of f. If f(t) = t, μ is the Lebesgue measure and the integral $\int_I g\, d\mu$ is written as

$$\int_a^b g(t)dt.$$

When the set E is regarded as a variable, $\int_E g\, d\mu$ is called the indefinite integral of g with respect to μ. It is of bounded variation and absolutely continuous with respect to μ.

Lemma 1.1 Let (X, Σ, μ) be a measure space and f a complex-valued μ-integrable function and

$$\lambda(E) = \int_E f\, d\mu \quad , \quad E \in \Sigma.$$

Then a function g on X to a Banach space Y is λ-integrable if and only if fg is μ-integrable and in this case the relation

$$\int_E g\, d\lambda = \int_E f g\, d\mu \quad , \quad E \in \Sigma$$

holds.

1.3 Distribution and Distributional Derivative.

Let X be a topological space. The support of a complex function f on X is the closure of the set

$$\{ x \in X ; f(x) \neq 0 \}.$$

For an open subset Ω of R^n, we denote by $C_c^\infty(\Omega)$, the collection of all infinitely partially differentiable functions defined on Ω and having a compact support. A classical example of a function

$\in C_c^\infty(R^n)$ is

$$\phi(x) = \begin{cases} \exp[(\|x\|^2-a^2)^{-1}], & \text{for } \|x\| = (\sum_{i=1}^{n} |x_i|^2)^{1/2} < a,\ a > 0 \\ 0 & , \text{for } \|x\| \geq a. \end{cases}$$

$C_c^\infty(\Omega)$ is a normed linear space under addition, scalar multiplication and norm defined by

$$(\phi_1 + \phi_2)(x) = \phi_1(x) + \phi_2(x)$$

$$(\alpha\,\phi)(x) \qquad = \alpha\,\phi(x)$$

$$\|\phi\| = \sup_{x \in \Omega} |\phi(x)|.$$

A continuous linear functional on $C_c^\infty(\Omega)$ is called a distribution on Ω. The space of distributions on Ω, being the dual space of $C_c^\infty(\Omega)$, is denoted by $C_c^\infty(\Omega)'$.

If μ is a complex Borel measure on Ω, then

$$T_\mu(\phi) = \int_\Omega \phi\, d\mu \quad , \quad \phi \in C_c^\infty(\Omega)$$

defines a distribution on Ω. In fact, by Riesz Representation Theorem, the set of all complex Borel measures on Ω is, by $\mu \longleftrightarrow T$, in one - one correspondence with the set of all distributions on Ω.

Let a complex function f, defined almost everywhere on Ω, be locally integrable on Ω with respect to the Lebesgue measure, in the sense that $\int_K |f(x)|dx < \infty$ for each compact subset K of Ω. Then

$$T_f(\phi) = \int_\Omega f(x)\,\phi(x)dx, \quad \phi \in C_c^\infty(\Omega) \tag{1.1}$$

defines a distribution on Ω. Two distributions T_f and T_g are equal as functionals (that is $T_f(\phi) = T_g(\phi)$ for every $\phi \in C_c^\infty(\Omega)$) if and only if $f(x) = g(x)$ a.e. on Ω. Hence, the set of all locally integrable functions on Ω is, by $f \longleftrightarrow T_f$, in one-one correspondence with a subset of $C_c^\infty(\Omega)'$ in such a way that (f and g being considered equivalent if and only if $f(x) = g(x)$ a.e.),

$$T_f + T_g = T_{f+g} \text{ , } \alpha T_f = T_{\alpha f}.$$

The derivative of a distribution T with respect to x_i, denoted by $D_i T$ (or $\partial T / \partial x_i$), is defined by

$$D_i T = - T(\frac{\partial \phi}{\partial x_i}), \quad \phi \in C_c^\infty(\Omega)$$

and is also a distribution on Ω. A distribution is infinitely differentiable in the sense of above definition.

Since a locally integrable function f on an open interval I of the real line can be identified with the distribution T_f on I, $D T_f$ ($\equiv d T_f / dt$) is denoted by $D T$ and is called the distributional derivative of f to distinguish it from the ordinary derivative $f'(\equiv df / dt)$. If f is absolutely continuous, then Df is the ordinary derivative f' (which is defined a.e.), f' being identified with the distribution $T_{f'}$. If f is of bounded variation, then Df is the Lebesgue-Stieltjes measure df, df being identified with the distribution T_{df}. Thus, for the Heaviside function $H(t)$ defined by

$$H(t) = \begin{cases} 0 & \text{if } t < 0 \\ 1 & \text{if } t \geq 0 \end{cases},$$

we have

$$D H \equiv D T_H \equiv T_\delta \equiv \delta ,$$

where δ is the Dirac measure. Indeed, for any $\phi \in C_c^\infty(R)$, we have

$$D T_H(\phi) = - T_H(\frac{d\phi}{dt}) = - \int_{-\infty}^{\infty} H(t) \phi'(t)dt$$

$$= - \int_0^\infty \phi'(t)dt = \phi(0) ,$$

since ϕ has compact support. Note that the ordinary derivative of H is the zero function a.e. on R.

Now consider the measure differential equation

$$Dx = F(t,x) + G(t,x)Du \tag{1.2}$$

where F and G are defined on $J \times R^n$ with values in R^n and u

is a right-continuous function $\in BV(J,R)$. Let S be an open connected set in R^n and I an interval with left end point $T_0 \geq t_0$.

Definition 1.4. A function $x(.) = x(.; T_0, x_0)$ is said to be a solution of (1.2) through (T_0, x_0) on I if $x(.)$ is a right-continuous function $\in BV(I,S)$, $x(T_0) = x_0$ and the distributional derivative of $x(.)$ on (T_0, τ) for any arbitrary $\tau \in I$ satisfies (1.2).

For example, the solution of

$$Dx = 1 + DH , \quad x(o) = 0$$

where H is the Heaviside function, is

$$x(t) = \begin{cases} t - 1 & \text{if } t < 0 \\ t & \text{if } t \geq 0. \end{cases}$$

1.4 Growth Problem

Growth of bacteria in medicine, decay of radioactive elements in physics, increase in population and pollution etc. are some of the problems studied by employing differential equations of evolution type.

Let $x(t)$ denote the quantity of an entity under consideration at time t. Assuming that the rate of growth of $x(t)$ at any time t is proportional to $x(t)$, we arrive at a simple differential equation

$$(1) \qquad \frac{dx(t)}{dt} = \alpha\, x(t), \quad x(t_0) = x_0 , \qquad (\alpha \text{ is real})$$

where α is the constant of proportionality. The growth or decay of the entity would depend on the sign of α. The unique solution of this problem is then given by

$$(2) \qquad x(t) = e^{\alpha(t - t_0)} x_0 , \qquad t \geq t_0 .$$

Let us study a somewhat new situation where this model needs to be modified.

Consider a fish breeding pond where in fish are grown scientifically. Some variety of fish breed is released in a pond and allowed to grow. After some fixed time intervals $t_1, t_2 \ldots$ partially grown fish are removed from the pond and simultaneously new breed of fish is released in it. The growth of fish population is impulsive. The impulses are given at times $t_1, t_2 \ldots$. This problem can be solved satisfactorily if the model (1) is replaced by

$$(3) \qquad Dx(t) = \alpha\, x(t)\, Du \;, \quad x(t_o) = x_o \;;$$

where D is the distributional derivative, u is a right continuous function of bounded variation.

Assume that u is of the form $u(t) = t + \sum_{k=1}^{\infty} a_k H_k(t)$

$$\text{where } H_k(t) = \begin{cases} 0 & \text{if } t < t_k \\ 1 & \text{if } t \geq t_k \end{cases}$$

where a_k are real numbers. Generally a right continuous function of bounded variation contains an absolutely continuous part and a singular part. Note that the discontinuities of u are isolated. Further

$$Du = 1 + \sum_{k=1}^{\infty} a_k H_k(t_k)$$

where $H_k(t_k)$ is the Dirac measure concentrated at t_k. It has been shown subsequently that the unique solution of this equation is given by

$$(4) \qquad x(t) = \frac{e^{\alpha(t - t_o)}}{\prod_{i=1}^{k-1} (1 - a_i)}\, x_o \;, \quad a_i \neq 1, \quad t_{k-1} \leq t < t_k \;.$$

Clearly, if $a_i = 0$ for $i = 1,2,\ldots$ in (4) then the equation (3) reduces to (1) and the solution (4) reduces to (2).

By considering several varieties of fish growing in one pond, the model of growth given in (3) can be generalized to a system of equations.

Before we proceed further, we wish to bring to the notice of readers the following important point : The terms distribution and distributional derivative used in the monograph are in a special sense and not in the general sense, as for example widely used in the literature on partial differential equations. Here, by a distribution, we shall always mean a continuous linear functional generated by a locally integrable function, given by formula (1.1). With this understanding, it follows that the measure differential equation (1.2) is identified with the ordinary differential equation $x' = F(t,x) + G(t,x)\, u'$ when u is an absolutely continuous function.

1.5 Notes.

Theorem 1.1 is known as Helly's Selection Principle, its proof is given in Graves [11, Chap. XII, Theorem 33]. Lemma 1.1 is taken from Dunford and Schwartz [8, Cor.6, p.180]. The approach to the Theory of distributions used in this monograph is due to Halperin [13], see also Yosida [51] for more details. For an alternative approach to the study of differential systems with discontinuous solutions see Stallard [46], Halany and Wexler [12], Fleishman and Mahar [9], Ligeza[24,25], Stuart [48] and Mottoni and Texi [27].

CHAPTER 2

EXISTENCE AND UNIQUENESS

Consider the measure differential equation

$$Dx = F(t,x) + G(t,x)\,Du \tag{2.1}$$

where F and G are defined on $J \times R^n$, $J = [0, \infty)$, with values in R^n, u is a right-continuous function $\in BV(J,R)$ and Dx, Du denote the distributional derivatives of x and u respectively. u being a function of bounded variation, Du can be identified with a Stieltjes measure which has the effect of instantaneously changing the state of the system (2.1) at the points of discontinuity of u. In this chapter, we consider the problem of existence and uniqueness of (2.1). Let S be an open connected set in R^n and I an interval with left end point $t_0 \geq 0$. Suppose that for each $x(\cdot) \in BV(I,S)$, $F(t,x(t))$ is Lebesgue integrable and $G(t,x(t))$ is integrable with respect to the Lebesgue-Stieltjes measure du.

2.1 Integral Representation

Qualitative properties of solutions of differential equations can be best studied by converting the differential equation into an integral equation. It is a matter of trivial verification that the ordinary differential equation

$$x' = f(t,x), \quad x(t_0) = x_0$$

is equivalent to the integral equation

$$x(t) = x_0 + \int_{t_0}^{t} f(s,x(s))\,ds.$$

For measure differential equations also we have a similar but non-trivial result.

<u>Theorem 2.1.</u> $x(\cdot)$ is a solution of (2.1) through (t_0,x_0) on I if and only if $x(\cdot)$ satisfies the integral equation

$$x(t) = x_0 + \int_{t_0}^{t} F(s,x(s))ds + \int_{t_0}^{t} G(s,x(s))du(s), \quad t \in I. \quad (2.2)$$

(Here the integral $\int_{t_0}^{t} G(s,x(s))du(s)$ is considered over the semi-open interval $(t_0,t]$.)

The proof depends upon the following :

<u>Lemma 2.1.</u> If g is a function integrable with respect to μ and T is a distribution on Ω given by

$$T(\phi) = \int_{\Omega} \phi \, d\mu \quad , \phi \in C_c^{\infty}(\Omega)$$

then the product gT defined by

$$(gT)(\phi) = \int_{\Omega} g \phi d\mu \quad , \phi \in C_c^{\infty}(\Omega) \qquad (2.3)$$

is also a distribution on Ω.

<u>Proof</u>. Since $\phi \in C_c^{\infty}(\Omega)$, it is bounded and μ-measurable. g being μ-integrable, the product $g\phi$ is μ-integrable. Hence the right hand side of (2.3) is meaningful. Clearly gT is a linear functional on $C_c^{\infty}(\Omega)$. Moreover,

$$|(gT)(\phi)| \leq \int_{\Omega} |g| \, |\phi| \, d\,|\mu| \leq \|\phi\| \int_{\Omega} |g| \, d\,|\mu| \, ,$$

where $|\mu|$ is the total variation measure of μ. Therefore,

$$\|gT\| = \sup_{\|\phi\| \leq 1} \{|(gT)(\phi)|\} \leq \int_{\Omega} |g| \, d\,|\mu| < \infty \, ,$$

since μ-integrability of g implies $|\mu|$-integrability of $|g|$. Thus gT is a bounded (hence continuous) linear functional on $C_c^{\infty}(\Omega)$, and is therefore a distribution on Ω.

<u>Proof of Theorem 2.1.</u> Suppose $x(.)$ satisfies (2.2) for $t \in I$. The integral $\int_{t_0}^{t} F(s,x(s))ds$ is absolutely continuous (and

hence continuous and of bounded variation) function of t on I. The integral

$$\int_{t_0}^{t} G(s,x(s))du(s)$$

is a function of bounded variation on I, and moreover, it is a right-continuous function of t since u is so. Therefore, $x(.)$ is a right-continuous function $\in BV(I,S)$. Clearly $x(t_0) = x_0$. If τ is any arbitrary point of I and T^i is the distribution on $I_\tau=(t_0,\tau)$ to be identified with the ith component $x^i(.)$ of $x(.)$, then

$$T^i(\phi) = \int_{I_\tau} \{x_o^i + \int_{t_0}^{t} F^i(s,x(s))ds + \int_{t_0}^{t} G^i(s,x(s))du(s)\}\phi(t)dt \tag{2.4}$$

for all $\phi \in C_c^\infty(I_\tau)$. Therefore

$$\begin{aligned}(DT^i)(\phi) &= -T^i(\phi') \\ &= -\int_{I_\tau} \{x_o^i + \int_{t_0}^{t} F^i(s,x(s))ds + \int_{t_0}^{t} G^i(s,x(s))du(s)\}\phi'(t)dt .\end{aligned} \tag{2.5}$$

Now

$$\begin{aligned}&-\int_{I_\tau} \{x_o^i + \int_{t_0}^{t} F^i(s,x(s))ds\}\ \phi'(t)dt \\ &= \int_{I_\tau} \phi(t)\ F^i(t,x(t))dt ,\end{aligned} \tag{2.6}$$

by the theorem on integration by parts and the fact that $\phi(t_0) = 0 = \phi(\tau)$.

The function $g(t) = \int_{t_0}^{t} G^i(s,x(s))du(s)$ is right-continuous and of bounded variation on I_τ. We have

$$\begin{aligned}\int_{I_\tau} g(t)\ \phi'(t)\ dt &= \int_{(t_0,\tau)} g(t)d\ \phi(t) \\ &= \int_{(t_0,\tau]} g(t)d\ \phi(t) - \int_{\{\tau\}} g(t)d\ \phi(t).\end{aligned}$$

But $\int_{\{\tau\}} g(t)\, d\,\phi(t) = 0$, since ϕ is continuous, and

$$\int_{(t_0,\tau)} g(t)d\phi(t) = g(\tau)\phi(\tau)-g(t_0)\phi(t_0)- \int_{(t_0,\tau)} \phi(t)dg(t)$$

$$= - \int_{(t_0,\tau)} \phi(t)dg(t)$$

$$= - \int_{(t_0,\tau)} \phi(t)dg(t) - \int_{\{\tau\}} \phi(t)dg(t)$$

$$= - \int_{I_\tau} \phi(t)dg(t) - \phi(\tau)\, (g(\tau) - g(\tau-))$$

$$= - \int_{I_\tau} \phi(t)dg(t)$$

$$= - \int_{I_\tau} \phi(t)\, G^i(t,x(t))du(t) \ , \qquad (2.7)$$

by Lemma 1.1. From (2.5) - (2.7), we obtain

$$(DT^i)(\phi) = \int_{I_\tau} \phi(t)\, F^i(t,x(t))dt + \int_{I_\tau} \phi(t)\, G^i(t,x(t))du(t). \qquad (2.8)$$

The second continuous linear functional on the right hand side of (2.8) is, by Lemma 2.1, identified with the measure $G^i(t,x(t))du(t)$, while the first continuous linear functional there is identified with $F^i(t,x(t))$. Since this holds for each $i = 1,2,\ldots,n$, the derivative $DT(\phi)$ is identified with $F(t,x(t)) + G(t,x(t))Du$ and therefore, $x(.)$ is a solution of (2.1) through (t_0, x_0).

Conversely, suppose $x(.)$ is a solution of (2.1) through (t_0, x_0) on I. Then for $I_\tau = (t_0, \tau)$, where τ is an arbitrary point of I, we have

$$\int_{I_\tau} \phi(t)\, d\, x^i(t) = \int_{I_\tau} \phi(t)\, F^i(t,x(t))dt$$

$$+ \int_I \phi(t)\, G^i(t,x(t))du(t) \qquad (2.9)$$

for $i = 1,2,\ldots,n$ and for each $\phi \in C_c^\infty(I_\tau)$.

Integrating the left hand side of (2.9) by parts and using (2.6) and (2.7), we obtain

$$\int_{I_\tau} \phi'(t)(x^i(t) - x_o^i)dt$$

$$= \int_{I_\tau} \phi'(t) \left\{ \int_{t_0}^{t} F^i(s,x(s))ds + \int_{t_0}^{t} G^i(s,x(s))du(s)\right\}dt.$$

This implies

$$x^i(t) = x_o^i + \int_{t_0}^{t} F^i(s,x(s))ds + \int_{t_0}^{t} G^i(s,x(s))du(s) \qquad (2.10)$$

a.e. in I_τ. But, since $x^i(\cdot)$ is right-continuous, $x(\cdot)$ being solution of (2.1), and since the right hand side of (2.10) is a right-continuous function of t, equality holds everywhere in I_τ for (2.10) and thus $x(\cdot)$ satisfies (2.2) for $t \subset I$.

q.e.d.

2.2 Existence of Solutions.

Let S and I be as in Section 2.1. Define

$$S_b(x_0) = \{x \in R^n ;\ |x-x_0| < b\}$$

$$E = \{(t,x);\ t \in [t_0, \hat{t}],\ x \in S_b(x_0)\}$$

where $x_0 \in S$ and $b > 0$ is so small that $S_b(x_0) \subset S$.

<u>Theorem 2.2.</u> (Local Existence). Let F and G satisfy the following conditions on E :

(i) $F(t,x)$ is measurable in t for each x and continuous in x

for each t ;

(ii) there exists a Lebesgue integrable function f such that

$$|F(t,x)| \leq f(t) \quad \text{for} \quad (t,x) \in E \text{ ;}$$

(iii) $G(t,x)$ is continuous in x for each t and $G(t,x(t))$ is du-integrable for each $x(.) \in BV([t_o,\hat{t}], S_b(x_o))$;

(iv) there exists a dv_u-integrable function g such that

$$|G(t,x)| \leq g(t) \quad \text{for} \quad (t,x) \in E \text{ ,}$$

where v_u denotes the total variation function of u.

Then there exists a solution $x(.)$ of (2.1) on some interval $[t_o, t_o + a]$ satisfying the initial condition $x(t_o) = x_o$.

<u>Proof</u>. Choose a, $0 < 2a < \hat{t} - t_o$ such that

$$\int_{t_o}^{t_o+2a} f(s)ds + \int_{t_o}^{t_o+2a} g(s)dv_u(s) \leq c \tag{2.11}$$

where $0 < c < b$. Such a choice of a is possible since

$$\int_{t_o}^{t} f(s)ds \quad \text{and} \quad \int_{t_o}^{t} g(s)dv_u(s)$$

are respectively continuous and right-continuous function of t.

Consider the following integral equation

$$x^{(k)}(t) = \begin{cases} x_o \ , & \text{for} \quad t \in [t_o - 2\,a/k,\ t_o] \\ x_o + \int_{t_o}^{t} F(s,x^{(k)}(s-2a/k))ds + \int_{t_o}^{t} G(s,x^{(k)}(s-2a/k))du(s), & \\ & \text{for } t \in (t_o,\ t_o + 2a]. \end{cases} \tag{2.12}$$

For any k, the first expression in (2.12) defines $x^{(k)}$ on $[t_o - 2a/k,\ t_o]$ and since $(t,\ x_o) \in E$ for $t \in [t_o,\ t_o + 2a/k]$, the second expression there defines $x^{(k)}$ as a function of bounded variation on $(t_o, t_o + 2a/k]$. Suppose that $x^{(k)}$ is defined on

$[t_0 - 2a/k,\ t_0 + 2aj/k]$, $1 \leq j < k$. Then for $t \in [t_0-2a/k, t_0+2aj/k]$, we have

$$|x^{(k)}(t) - x_0| \leq V(x^{(k)}(.) - x_0,\ [t_0 - 2a/k,\ t_0 + 2aj/k])$$

$$\leq \int_{t_0}^{t_0+2aj/k} |F(s,x^{(k)}(s-2a/k))|ds + \int_{t_0}^{t_0+2aj/k} |G(s,x^{(k)}(s-2a/k))| dv_u(s)$$

$$\leq c < b \tag{2.13}$$

by (2.11), so that the second expression in (2.12) defines $x^{(k)}$ on $(t_0 + 2aj/k,\ t_0 + 2a(j+1)/k]$. Thus $x^{(k)}$ is defined on $[t_0 - 2a/k,\ t_0 + 2a]$, and as in (2.13) it follows that

$$|x^{(k)}(t) - x_0| \leq V(x^{(k)}(.) - x_0, [t_0 - 2a/k,\ t_0 + 2a]) < b. \tag{2.14}$$

Define

$$\tilde{x}^{(k)}(t+2a/k) = x^{(k)}(t),\ t_0 - 2a/k \leq t \leq t_0 + 2(1-1/k)a,\ k \geq 2.$$

Then $\tilde{x}^{(k)}$ is right continuous, since $x^{(k)}$ is so. Further by (2.14), $\tilde{x}^{(k)}$ are of uniform bounded variation on $[t_0, t_0+a]$. By Theorem 1.1, there is a subsequence $\tilde{x}^{(kj)}$ and a function x^* such that

$$\lim_{j \to \infty} \tilde{x}^{(k_j)}(t) = x^*(t),\ t \in [\ t_0,\ t_0 + a\ ]. \tag{2.15}$$

Continuity of F and G in x implies

$$\lim_{j \to \infty} F(t,\ \tilde{x}^{(k_j)}(t)) = F(t,\ x^*(t)), \tag{2.16}$$

$$\lim_{j \to \infty} G(t,\ \tilde{x}^{(k_j)}(t)) = G(t, x^*(t)). \tag{2.17}$$

Therefore, Lebesgue dominated convergence theorem gives

$$\lim_{j \to \infty} \int_{t_0}^{t} F(s,\ \tilde{x}^{(k_j)}(s))ds = \int_{t_0}^{t} F(s, x^*(s))ds$$

and

$$\lim_{j\to\infty} \int_{t_o}^{t} G(s,\tilde{x}^{(k_j)}(s))du(s) = \lim_{j\to\infty} \{ \int_{t_o}^{t} G(s,\tilde{x}^{(k_j)}(s))du^+(s) - \int_{t_o}^{t} G(s,\tilde{x}^{(k_j)}(s))du^-(s) \}$$

$$= \int_{t_o}^{t} G(s,x^*(s))du^+(s) - \int_{t_o}^{t} G(s,x^*(s))du^-(s)$$

$$= \int_{t_o}^{t} G(s,x^*(s))du(s),$$

where du^+ and du^- respectively denote the positive and negative variations of the Lebesgue-Stieltjes measure du.

From (2.12), we have

$$\tilde{x}^{(k_j)}(t+2a/k_j) = \begin{cases} x_o, & \text{for } t \in [t_o - 2a/k_j, t_o] \\ x_o + \int_{t_o}^{t} F(s,\tilde{x}^{(k_j)}(s))ds + \int_{t_o}^{t} G(s,\tilde{x}^{(k_j)}(s))du(s), & \text{for } t \in (t_o, t_o + a]. \end{cases}$$

Taking limit as $j \to \infty$, we obtain

$$x^*(t) = x_o + \int_{t_o}^{t} F(s,x^*(s))\, ds + \int_{t_o}^{t} G(s,x^*(s))du(s),$$

for $t \in [t_o, t_o + a]$. Hence, by Theorem 2.1, x^* is a solution of (2.1) through (t_o, x_o) on $[t_o, t_o + a]$.

q. e. d.

2.3 Uniqueness

The fact that u may have discontinuities offers difficulties in establishing the uniqueness of solutions of (2.1). In fact the following simple example shows that the solutions of (2.1) through a given point need not be unique even when F and G are linear in x (see also Example 4.1).

<u>Example 2.1.</u> Consider the scalar equation

$$Dx = 2(t+1)^{-1}\, xDu, \quad t \in [0,2] \tag{2.18}$$

where

$$u(t) = \begin{cases} t, & 0 \leq t < 1 \\ t+1, & 1 \leq t \leq 2 \end{cases}.$$

From Theorem 2.1, it is easy to see that there are infinitely many solutions of (2.18) through (0,0) on [0,2], given by

$$x(t) = \begin{cases} 0, & 0 \leq t < 1 \\ \dfrac{c(t+1)^2}{4}, & 1 \leq t \leq 2 \end{cases},$$

where c is an arbitrary constant.

However, if u is of type $\mathcal{Y}$, then under certain conditions, it is possible to obtain uniqueness. To see this, we first prove

<u>Theorem 2.3.</u> Suppose that

(i) x and g are nonnegative functions, defined and du-integrable on J;

(ii) u is of type $\mathcal{Y}$; u' is non-negative on $J - \{t_k\}$; the jumps $a_k = u(t_k) - u(t_k-)$ of u at $t = t_k$ are such that

$$|a_k|\, g(t_k) < 1, \quad k = 1,2,\ldots \tag{2.19}$$

and

$$\sum_{k=1}^{\infty} |a_k|\, g(t_k) < \infty. \tag{2.20}$$

Then, for any positive constant c, the inequality

$$x(t) \leq c + \int_{t_0}^{t} g(s)x(s)\, du(s), \quad t \in J \tag{2.21}$$

implies

$$x(t) \leq c\, P^{-1} \exp\left(\int_{t_0}^{t} g(s)\, u'(s)\, ds\right), \quad t \in J \tag{2.22}$$

where

$$P = \prod_{k=1}^{\infty} (1 - |a_k|\ g(t_k)). \tag{2.23}$$

Proof. Denote the quantity on the right hand side of (2.21) by $r(t)$. Since u is differentiable on $[t_0, t_1)$, Gronwall's inequality gives

$$r(t) \leq c \exp(\int_{t_0}^{t} g(s)u'(s)\, ds),\ t \in [t_0, t_1).$$

At $t = t_1$, we have, for sufficiently small $h > 0$,

$$r(t_1) = r(t_1-h) + \int_{t_1-h}^{t_1} g(s)x(s)\, du(s)$$

$$\leq c \exp(\int_{t_0}^{t_1-h} g(s)u'(s)ds) + \int_{t_1-h}^{t_1} g(s)r(s)du(s). \tag{2.24}$$

We prove that

$$\lim_{h \to 0+} |\int_{t_1-h}^{t_1} g(s)r(s)du(s)| = |a_1|\ g(t_1)r(t_1). \tag{2.25}$$

Consider the set function λ defined by

$$\lambda(A) = |\int_A g(s)\ r(s)\ du(s)|.$$

Let $\{h_k\}$ be a decreasing sequence of positive real numbers tending to zero and let $A_k = [t_1 - h_k, t_1]$. Then $A_1 \supset A_2 \supset \dots$ and $\bigcap_{k=1}^{\infty} A_k = \{t_1\}$. Therefore, $\lambda(A_k) \to \lambda(\{t_1\}) = |a_1|\ g(t_1)r(t_1)$, and (2.25) is established. Taking limit as $h \to 0_+$ in (2.24), we obtain

$$r(t_1) \leq c \exp(\int_{t_0}^{t_1} g(s)u'(s)ds + |a_1|\ g(t_1)r(t_1),$$

which, in view of (2.19), yields

$$r(t_1) \leq c\ P_1^{-1} \exp(\int_{t_0}^{t_1} g(s)\ u'(s)\ ds), \tag{2.26}$$

where $P_k = \prod_{i=1}^{k} (1 - |a_i|\ g(t_i))$.

For $t \in [t_1, t_2)$, we note that

$$r(t) = r(t_1) + \int_{t_1}^{t} g(s)\ x(s)du(s)$$

$$\leq r(t_1) + \int_{t_1}^{t} g(s)\ r(s)du(s).$$

As above, this gives

$$r(t) \leq r(t_1) \exp\ (\int_{t_1}^{t} g(s)u'(s)ds)$$

$$\leq c\ P_1^{-1}\exp(\int_{t_0}^{t} g(s)u'(s)ds),\ t \in [t_0,t_2),$$

from (2.26). By induction,

$$r(t) \leq c\ P_{k-1}^{-1} \exp\ (\int_{t_0}^{t} g(s)u'(s)ds),\ t \in [t_0,t_k). \tag{2.27}$$

From (2.19) and (2.20), we observe that $P_k \geq P_{k+1}$ for each k and $\lim_{k\to\infty} P_k = P$. Therefore, for any $t \in J$, (2.27) implies

$$x(t) \leq r(t) \leq c\ P^{-1}\exp\ (\int_{t_0}^{t} g(s)u'(s)ds).$$

q. e. d.

Example 2.2. Consider the inequality

$$x(t) \leq c + \int_{1}^{t} s^{-3}x(s)\ du(s),\ c > 0,\ t \in J = [1,\infty)$$

where

$$u(t) = t^2 + \sum_{k=2}^{\infty} (\sum_{i=2}^{k} i)\chi_{[k-1,k)}(t).$$

Here χ_A denotes the characteristic function of the set A. Clearly u is of the type $\mathcal{S}$ with $t_k = k$, $k = 2,3,\ldots$. Further, $a_k = k$,

$|a_k|\ g(t_k) = k^{-2} < 1$ for each $k \geq 2$, $\sum_{k=2}^{\infty} k^{-2} < \infty$ and $P = 1/2$.
Thus, all the conditions of Theorem 2.3 are satisfied. Estimate (2.22) gives

$$x(t) \leq 2c \exp\ (2(1-t^{-1})),\ \text{for all } t \in J.$$

Example 2.3. Let $J = [0,\infty)$. In the inequality (2.21) choose $g(t) = (t+1)^{-1}$ and

$$u(t) = \log(t+1) + \sum_{k=1}^{\infty} \left(\sum_{i=1}^{k} (i+1)^{-1} \right) \chi_{[k-1,k)}(t).$$

Then, all the conditions of Theorem 2.3 are satisfied and we have

$$x(t) \leq 2\ c \exp\ (t(t+1)^{-1}) \quad \text{for all } t \in J.$$

Note that, in Example 2.2, the series $\sum_{k=2}^{\infty} g(t_k)$ converges, whereas in Example 2.3 it diverges. However, in both the cases, the series $\sum_{k=2}^{\infty} |a_k|\ g(t_k)$ converges. Thus, Theorem 2.3 is applicable to system containing large as well as small impulses.

The following corollary is an immediate consequence of Theorem 2.3 and is useful in studying perturbed systems.

Corollary 2.1. If f is a non-negative function defined and integrable on J, then under the conditions of Theorem 2.3, the inequality

$$x(t) \leq c + \int_{t_0}^{t} f(s)x(s)ds + \int_{t_0}^{t} g(s)x(s)du(s),\ t \in J$$

implies

$$x(t) \leq c\ P^{-1}\exp\ \left(\int_{t_0}^{t} [f(s)+g(s)\ u'(s)]ds \right),\ t \in J.$$

Uniqueness of solutions of (2.1) is now easy to obtain. Note that, if u is of type $\mathcal{Y}$, so is its total variation function v_u. Moreover v_u is non-decreasing on (t_{k-1}, t_k) for each k, so that $v_u' \geq 0$ on $J - \{t_k\}$. Let $b_k = v_u(t_k) - v_u(t_k-)$, $k = 1,2,\ldots$.

<u>Theorem 2.4.</u> Suppose that

(i) u is of type $\mathcal{Y}$;

(ii) there exist functions f and g on J such that

$$|F(t,x)-F(t,y)| \leq f(t)\,|x-y|, \quad |G(t,x)-G(t,y)| \leq g(t)|x-y|$$

for $t \in J$ and $|x|, |y| \leq H$, $H > 0$;

(iii) b_k satisfies

$$|b_k|\, g(t_k) < 1, \quad k = 1,2,\ldots \tag{2.28}$$

and

$$\sum_{k=1}^{\infty} |b_k|\, g(t_k) < \infty, \quad Q = \prod_{k=1}^{\infty} (1 - |b_k|\, g(t_k));$$

(iv) $f,\ g\ v_u' \in L^1(J).$ (2.29)

Then (2.1) has at the most one solution through (t_0, x_0).

<u>Proof.</u> For any two solutions $x(t) = x(t,t_0,x_0)$ and $y(t) = y(t, t_0, x_0)$ of (2.1) and for any $\varepsilon > 0$, we have from Theorem 2.1 and condition (i)

$$z(t) < \varepsilon + \int_{t_0}^{t} f(s)z(s)ds + \int_{t_0}^{t} g(s)z(s)dv_u(s), \quad t \in J$$

where $z(t) = |x(t) - y(t)|$. Corollary 2.1 yields

$$z(t) \leq \varepsilon\, Q^{-1} \exp\left(\int_{t_0}^{t} [f(s)+g(s)\, v_u'(s)]ds \right), \quad t \in J.$$

The desired conclusion now follows from (2.29), since $\varepsilon > 0$ is arbitrary.

q. e. d.

Note that condition (2.28) fails to hold in Example 2.1, since at the point of discontinuity $t = 1$ there, we have $b_1 = -1$ so that $|b_1|\, g(1) = 1$.

<u>Example 2.4.</u> Consider the scalar equation

$$Dx = e^{-t}\, x^3 + \frac{e^{-2t}}{2}\, (t^3+1)^{-1}\, x^2\, Du, \quad t \in [1,\infty), \ |x| \leq 1$$

where $u(t) = e^t + \sum_{k=2}^{\infty} \left(\sum_{i=1}^{k-1} e^i \right) \chi_{[k-1,k)}(t)$. Then

$$v_u(t) = e^t - \sum_{k=2}^{\infty} e^{k-1} \chi_{[k-1,k)}(t), \quad t_k = k, \ b_k = e^{k-1} - e^k$$

for $k = 2,3,\ldots$. If we choose $f(t) = 3\, e^{-t}$, $g(t) = e^{-t}(t^3+1)$, then all the conditions of Theorem 2.4 are satisfied.

According to Definition 1.4, solution of a measure differential equation does not necessarily depend continuously on time but is instead a function of bounded variation in t. The following example shows that a solution need not depend continuously on the initial time t_o either.

<u>Example 2.5.</u> Consider the equation

$$Dx = x + Du \tag{2.30}$$

where $u(t) = \begin{cases} 0 & , t < 0 \\ 1 & , t \geq 0 \end{cases}$, so that Du is the Dirac measure.

For $\tau_1 < 0 < \tau_2$, find solutions of (2.30) through $(\tau_1, 1)$ and $(\tau_2, 1)$ respectively. Indeed, these are given by (see Theorem 3.1 of Chapter 3)

$$x(t, \tau_1, 1) = e^{t-\tau_1} + \int_{\tau_1}^{t} e^{t-s}\, du(s),$$

$$x(t, \tau_2, 1) = e^{t-\tau_2} + \int_{\tau_2}^{t} e^{t-s}\, du(s).$$

Hence

$$x(t, \tau_1, 1) - x(t, \tau_2, 1) = e^t \left[e^{-\tau_1} - e^{-\tau_2} - \int_{\tau_2}^{t} e^{-s} du(s) \right],$$

since $\int_{\tau_1}^{t} e^{-s}\, du(s) = 0$, if $t < 0$, using integration by parts,

$$- \int_{\tau_2}^{t} e^{-s}\, du(s) = - \int_{\tau_2}^{t} u(s)\, e^{-s}\, ds + e^{-\tau_2}$$

$$= \int_{0}^{\tau_2} e^{-s}\, ds + e^{-\tau_2} = 1.$$

Therefore,

$$x(t, \tau_1, 1) - x(t, \tau_2, 1) = e^{t}[e^{-\tau_1} - e^{-\tau_2} + 1].$$

In particular, for values of t, τ_1 and τ_2 sufficiently near zero but such that $t < 0$ and $\tau_1 < 0 < \tau_2$, we see that $|x(t,\tau_1,1)-x(t,\tau_2,1)|$ is approximately equal to one.

2.4 An Optimal Control Problem

In ordinary optimal control problem, one is given an ordinary differential equation

$$\frac{dx}{dt} = f(t,x,u) \tag{2.31}$$

which is a mathematical model of some physical process. The problem of control is to select $u(t)$ (called control variable) on an interval of time $t_0 \leq t \leq t_1$ such that the solution $x(t)$ of (2.31) behaves in a prescribed manner on $[t_0, t_1]$. The usual prescribed behaviour consists of requiring that the $u(t)$ be found such that the solution $x(t)$ moves from some given initial point x_0 to a prescribed moving target $T(t)$ so as to minimize some cost function or performance index.

In this section, the control problem for the optimal control of a nonlinear dynamical system is defined. The system under consideration is described by measure differential equations, which depend on certain parameters called control variables.

Let

$I = [t_o, \beta]$, $t_o \geq 0$, $\beta > t_o$ be a fixed interval ;

S = an open connected set in R^n;

U = the set of all right-continuous functions $\in$ BV(I,Q) where Q is a nonempty compact subset of R.

Consider the measure differential equation

$$Dx = f(t,x(t),\ u(t)) + g(t,x(t),u(t))Du \tag{2.32}$$

satisfying

(A_1) $f, g : I \times S \times U \to R^n$ are continuous;

(A_2) there exists a Lebesgue integrable function $r(t)$ on I such that

$$|\ f(t,x,u)\ | \leq r(t)$$

uniformly with respect to $x \in S$ and $u \in U$;

(A_3) there exists a constant K such that

$$|\ g(t,x,u)\ | \leq K$$

uniformly with respect to $x \in S$ and $u \in U$.

Under these assumptions, (2.32) has a solution

$x(t) = x(t,\ t_o,\ x_o)$, given by

$$x(t) = x_o + \int_{t_o}^{t} f(s,x(s),u(s))ds + \int_{t_o}^{t} g(s,x(s),u(s))du(s) \tag{2.33}$$

which is a function of bounded variation for each choice of $u \in U$ and $x_o \in S$.

Definition 2.1. If $u \in U$ defined on the interval $[t_o,\ t_1] \in I$ is such that the corresponding solution $x(t)$ of (2.32) is also defined on $[t_o,\ t_1]$, the u is called an admissible control for the equation (2.32), and $x(t)$ is called the corresponding trajectory.

Definition 2.2. For a given real-valued continuous function defined on $I \times S \times Q$, the cost functional $C(u)$ of a control $u(t)$ on $[t_o,\ t_1]$ for the system (2.32) is defined by

$$C(u) = \int_{t_o}^{t_1} F(t,x(t),u(t))dt.$$

If $F(t,x,u) \equiv 1$, then $C(u) = t_1 - t_o$, the time duration over which the control is exerted. Such a cost function is called <u>time optimal</u>.

<u>Definition 2.3.</u> A <u>target set</u> is a family $\mathcal{T}$ of nonempty compact sets $T(t) \subset R^n$ defined for $t \in I$. We say that an admissible control $u(t)$ defined on $[t_o, t_1]$ transfers the initial point $x_o \in S$ to the target set $\mathcal{T}$, if the trajectory $x(t) = x(t,t_o,x_o,u)$ corresponding to $u(t)$ satisfies the relation $x(t_1) \in T(t_1)$.

Let A denote the set of all admissible controls which transfer x_o to $\mathcal{T}$.

<u>Definition 2.4.</u> A control $u^* \in A$ is called an optimal (minimal) control if

$$C(u^*) \leq C(u) \quad \text{for each} \quad u \in A.$$

<u>Theorem 2.5.</u> Given the control problem with the following data :

(a) equation (2.32) with f and g satisfying the assumptions (A_1) - (A_3) ;

(b) the initial point $x_o \in S$;

(c) a target set $\mathcal{T}$ of nonempty compact sets $T(t) \subset R^n$ defined on I and upper semicontinuous with respect to inclusion ;

(d) the set A of admissible controls $u(t)$ defined on subintervals $[t_o,t_1] \subset I$ with the same left end point (and perhaps different right end point $t_1 > t_o$) which transfer x_o to $\mathcal{T}$, and which is such that for all $u \in A$,

$$|\Delta u| \leq \Delta h$$

on each subinterval of $[t_o,t_1]$ where h is a given nondecreasing right-continuous function defined on $[t_o,t_1]$ (here, the symbol Δh on $[t_o,t_1]$ denotes $h(t_1) - h(t_o)$) ;

(e) the cost functional

$$C(u) = \int_{t_0}^{t_1} F(t,x(t),u(t))dt,$$

where F is a continuous real function defined on $I\times S\times Q$.

Then, if A is nonempty, there exists an optimal control in A.

<u>Proof.</u> By (d), each $u \in A$ is of uniform bounded variation. From (2.33), (A_2), (A_3) and the condition (d) of the theorem, we have

$$V(x,[t_0,t_1]) \leq \int_{t_0}^{t_1} r(s)ds + K\,(h(\beta) - h(t_0)) ,$$

which implies that the corresponding trajectories are also uniformly bounded.

Now, since A is nonempty, and the corresponding trajectories are uniformly bounded,

$$\inf \{C(u) : u \in A\} = \tilde{m} > -\infty.$$

Either A is a finite set in which case the theorem is trivially true, or we can select a sequence $\{u^{(k)}\}$ from A such that $u^{(k)}$ are defined on intervals $[t_0, t_1^{(k)}]$ for which $\{C(u^k)\}$ decreases monotonically to $\tilde{m}$. Select a subsequence (which we still label $\{u^{(k)}\}$) such that $t_1^{(k)} \to t_1^*$ monotonically. We consider the case when $\{t_1^{(k)}\}$ is monontonic decreasing. Choose $\hat{t}_1$ such that $\hat{t}_1 = t_1$ if $t_1^{(k)} = t_1$ for all k ; otherwise let $t_1^* < t_1^{(k_0+1)} \leq \hat{t}_1 < t_1^{(k_0)} < \beta$ for some k_0 (from now on all reference to the index k tacitly assumes $k > k_0$). Extend the controls $u^{(k)}$ to the interval $[t_0, \hat{t}_1]$ by defining

$$\hat{u}^{(k)}(t) = \begin{cases} u^{(k)}(t) & \text{if } t \in [t_0, t_1^{(k)}] \\ u^{(k)}(t_1^{(k)}) & \text{if } t \in (t_1^{(k)}, \hat{t}_1]. \end{cases} \tag{2.34}$$

Clearly $\hat{u}^{(k)}(t) \in Q$ for all $t \in [t_0, \hat{t}_1]$; $u^{(k)}$ is right-continuous on $[t_0, \hat{t}_1]$ and $|\Delta \hat{u}^{(k)}| \leq \Delta h$ on every subinterval of $[t_0, \hat{t}_1]$. Therefore, by Theorem 1.1, there exists a subsequence (again denoted by $\{\hat{u}^{(k)}\}$) and a function of bounded variation u^* such that

$$\lim_{k \to \infty} \hat{u}^{(k)}(t) = u^*(t) \tag{2.35}$$

everywhere on $[t_0, \hat{t}_1]$. Our aim is to show that $u^* \in A$ and is an optimal control.

(i) u^* is right-continuous :

Let $\tau \in [t_0, \hat{t}_1]$. Since $| \Delta \hat{u}^{(k)} | \leq \Delta h$ on every subinterval of $[t_0, \hat{t}_1]$ and since h is right-continuous at τ, for any $\varepsilon > 0$, there exists a $\delta > 0$ such that

$$|\hat{u}^{(k)}(\tau+s)- \hat{u}^{(k)}(\tau)| \leq h(\tau+s) - h(\tau),\ \tau \leq \tau + s \leq \hat{t}_1$$

$$< \varepsilon\ , \quad \text{for } 0 \leq s \leq \delta .$$

for any k. Therefore,

$$\lim_{s \to 0_+} \hat{u}^{(k)}(\tau + s) = \hat{u}^{(k)}(\tau) \quad \text{uniformly for } k.$$

Also

$$\lim_{k\to\infty} \hat{u}^{(k)}(\tau + s) = u^*(\tau + s).$$

Hence, $u^*(\tau + 0) = u^*(\tau)$. Since $\tau \in [t_0, \hat{t}_1)$ was arbitrary, u^* is right-continuous on $[t_0, \hat{t}_1)$.

Since $\hat{u}^{(k)}(t) \in Q$ for $t \in [t_0, \hat{t}_1]$, it follows that $u^*(t) \in Q$ for $t \in [t_0, \hat{t}_1]$. It is easy to see that

$$| \Delta u^* | \leq \Delta h \tag{2.36}$$

on every subinterval of $[t_0, t_1]$.

Let $x^{(k)}$ be the trajectory defined on $[t_0, t_1^{(k)}]$ corresponding to the control $u^{(k)}$. The extended control $\hat{u}^{(k)}$ coincides with $u^{(k)}$ on $[t_0, t_1^{(k)}]$ and it is continuous on $[t_1^{(k)}, \hat{t}_1]$. Under the given conditions, $x^{(k)}$ can be extended to $\hat{x}^{(k)}$ on $[t_0, t_1]$ and the extension is given by

$$\hat{x}^{(k)}(t) = x_0 + \int_{t_0}^{t} f(s, \hat{x}^{(k)}(s), \hat{u}^{(k)}(s))\,ds$$

$$+ \int_{t_0}^{t} g(s,\hat{x}^{(k)}(s),\hat{u}^{(k)}(s))\,d\hat{u}^{(k)}(s) \qquad (2.37)$$

for $t \in [t_0, \hat{t}_1]$. Since

$$V(\hat{u}^{(k)}, [t_0, \hat{t}_1]) = V(u^{(k)}, [t_0, t_1^{(k)}]),$$

total variations of $\hat{x}^{(k)}$ are uniformly bounded. Further, $\hat{x}^{(k)}$ are also uniformly bounded. Hence there exists a subsequence (again labeled $\hat{x}^{(k)}$) and a function x^* such that

$$\lim_{k \to \infty} \hat{x}^{(k)}(t) = x^*(t) \quad \text{on} \quad [t_0, \hat{t}_1]. \qquad (2.38)$$

By selecting the corresponding subsequence from $\{\hat{u}^{(k)}\}$, any of the preceding limit operations satisfied by $\hat{u}^{(k)}$ are not changed. Therefore, from (2.35) - (2.38), hypotheses (A_1) - (A_3) and Lebesgue's dominated convergence theorem, we obtain

$$x^*(t) = x_0 + \int_{t_0}^{t} f(s, x^*(s), u^*(s))ds$$

$$+ \int_{t_0}^{t} g(s,x^*(s), u^*(s))\,du^*(s), \; t \in [t_0, \hat{t}_1]. \qquad (2.39)$$

(ii) $$\lim_{k\to\infty} \hat{x}^{(k)}(t_1^{(k)}) = x^*(t_1^*) \quad : \qquad (2.40)$$

We have

$$|\hat{x}^{(k)}(t_1^{(k)}) - x^*(t_1^*)| \leq |\hat{x}^{(k)}(t_1^{(k)}) - x^*(t_1^{(k)})|$$
$$+ |x^*(t_1^{(k)}) - x^*(t_1^*)|. \qquad (2.41)$$

The second term on the right of (2.41) tends to zero as $k \to \infty$ since x is right-continuous (u^*being so). Further from integral representations (2.37), (2.39) and hypotheses (A_1) - (A_3), it follows after some computation that

$$|\, \hat{x}^{(k)}(t_1^{(k)}) - x^*(t_1^{(k)}) \,| \to 0 \quad \text{as } k \to \infty,$$

so that (2.40) is established.

(iii) u^* on $[t_0, t_1]$ belongs to A :

We have $\hat{x}^{(k)}(t_1^{(k)}) \in T(t_1^{(k)})$ for $k = 1,2,\ldots$ and

$$x^*(t_1^*) = \lim_{k\to\infty} x^{(k)}(t_1^{(k)}).$$

If $x^*(t_1^*)$ were not in $T(t_1^*)$, then there would exist a neighbourhood N of the (compact) set $T(t_1^*)$ such that $x(t_1^*)$ is not in the closure $\bar{N}$ of N. But by upper semicontinuity of $T(t)$, $T(t) \subset N$ for t sufficiently near t_1^*, so that $x^{(k)}(t_1^{(k)}) \in N$ for large k and yet $x^*(t_1^*) \in \bar{N}$. This contradiction shows that $x^*(t_1^*) \in T(t_1^*)$ and thereby establishes (iii).

(iv) u^* is optimal :

We have

$$C(\hat{u}^{(k)}) = \int_{t_0}^{t_1^{(k)}} F(s,\hat{x}^{(k)}(s), \hat{u}^{(k)}(s))ds$$

$$= \int_{t_0}^{t_1^*} F(s,\hat{x}^{(k)}(s), \hat{u}^{(k)}(s))ds + \int_{t_1^*}^{t_1^{(k)}} F(s,\hat{x}^{(k)}(s),\hat{u}^{(k)}(s))ds. \tag{2.42}$$

The second integral on the right of (2.42) approaches zero as $k \to \infty$ by uniform boundedness of $F(s, \hat{x}^{(k)}(s), \hat{u}^{(k)}(s))$ on $[t_0, t_1^{(k)}]$. Since

$$\hat{x}^{(k)}(t) \to x^*(t), \hat{u}^{(k)}(t) \to u^*(t)$$

everywhere on $[t_0, t_1^*]$, we have by Lebesgue's dominated convergence theorem

$$\lim_{k\to\infty} C(\hat{u}^{(k)}) = \int_{t_0}^{t_1^*} F(s,x^*(s), u^*(s))\, ds = C(u^*).$$

But $\lim_{k\to\infty} C(u^{(k)}) = \tilde{m}$. Hence $C(u^*) = \tilde{m}$, and u^* on $[t_0, t_1^*]$ is an optimal control.

So the theorem is proved when the sequence $\{t_1^{(k)}\}$ is monotone decreasing . With minor changes in the preceding argument, the case when $\{t_1^{(k)}\}$ is monotone increasing can be dealt with. The details are omitted.

q. e. d.

The following example illustrates a situation where the optimal control fails to exist.

Example 2.6. Consider the equations

$$D\, x_1 = x_2$$

$$D\, x_2 = - x_2 + u(t) + D\, u.$$

The initial point is $x_0 = \begin{bmatrix} 0 \\ 0 \end{bmatrix}$ and the target is the fixed point $\begin{bmatrix} 0 \\ 1 \end{bmatrix}$. The set $Q = [-1, 1]$ and A is the set of all controls of uniform bounded variation with fixed initial time $t = 0$. Consider the controls

$$u^{(k)}(t) = \begin{cases} 0 & \text{if } 0 \leq t < 1/k \\ 1 & \text{if } 1/k \leq t \end{cases}$$

defined on $[0, 1]$. Then

$$x_2(t) = x_2(0) + \int_0^t - x_2(s)ds + \int_0^t u(s)ds + u(t) - u(o)$$

and

$$x_2^{(k)}(t) = \begin{cases} 0 & \text{if } 0 \leq t < 1/k \\ 1 & \text{if } t = 1/k\ . \end{cases}$$

Hence

$$\int_0^{1/k} x_2(s)ds = 0 = x_1(1/k)\ ,$$

and the sequence $u^{(k)}(t)$ is a sequence of controls in A. But the infimum of the cost functional is

$$\lim_{k \to \infty} C(u^{(k)}) = \lim_{k \to \infty} 1/k = 0.$$

The only control $u^*(t)$ in A which will produce a minimal time of o is the multi-valued control $u(o) = 0$, $u(o) = 1$. Hence an optimal control does not exist.

Notice that there does not exist any nondecreasing function h(t) such that

$$|\Delta u^{(k)}| \leq \Delta h,$$

but that all other hypotheses of the theorem are satisfied.

2.5 Notes.

Sections 2.1 and 2.2 constitute the work of Das and Sharma [5]. Integral representation of measure differential equations (Theorem 2.1) has also been established by Schmaedeke [45], in which the integrals are Riemann - Stieltjes type. Theorem 2.3 is taken from Pandit [28] and Theorem 2.4 from Pandit [29]. Theorem 2.5 is a slight modification of Theorem 13 of Schmaedeke [45]. The optimal control problem for measure delay - differential equations has been investigated by Das and Sharma [6] and for pulse frequency modulation systems by Pavlidis [35].

CHAPTER 3

STABILITY AND ASYMPTOTIC EQUIVALENCE

One of the methods of approaching a nonlinear problem is to express the nonlinear function as the sum of a linear and a nonlinear function. The original problem can thus be viewed as a perturbation problem, in which the nonlinear function plays the role of perturbation. The solutions of the nonlinear (perturbed) system can then be compared with those of the linear (unperturbed) system. Here, we consider the linear ordinary differential system

$$x' = A(t)x \tag{3.1}$$

and the nonlinear measure differential systems

$$Dy = A(t)y + F(t,y)Du \tag{3.2}$$

$$Dz = A(t)z + F(t,z)Du + G(t,z)Du \tag{3.3}$$

where $x, y, z \in R^n$; $A(t)$ is an n by n matrix, continuous on $J = [0, \infty)$; F, G are defined on $J \times S$ (S an open set in R^n) with values in R^n and u is a right-continuous function $\in BV(J,R)$. We shall also consider the nonlinear ordinary differential system

$$x' = f(t,x) \tag{3.4}$$

and its perturbed measure differential system

$$Dy = f(t,y) + g(t,y)Du \tag{3.5}$$

3.1 Existence and Stability via Krasnoselskii's Fixed Point Theorem

We need the following result which expresses solutions of (3.2) in terms of solutions of (3.1).

<u>Theorem 3.1.</u> Let $X(t)$ denote a fundamental solution matrix of (3.1) with $X(0) = E$, the identity n by n matrix. If F is du-integrable, then any solution $y(t) = y(t,t_0,x_0)$ of (3.2) on J is

given by

$$y(t) = x(t) + \int_0^t X(t)\, X^{-1}(s)\, F(s,y(s))du(s) \tag{3.6}$$

where $x(t) = X(t)x_0$ is the solution of (3.1) through (t_0,x_0) on J.

Proof. In view of Theorem 2.1, it is enough to show that $y(t)$ given by (3.6) is a solution of the integral equation

$$z(t) = x_0 + \int_0^t A(s)z(s)ds + \int_0^t F(s,z(s))du(s). \tag{3.7}$$

Substitute $y(t)$ in (3.6) into the right hand side of (3.7) to obtain

$$x_0 + \int_0^t A(s)y(s)ds + \int_0^t F(s,y(s))du(s)$$

$$= x_0 + \int_0^t A(s)X(s)x_0 ds + \int_0^t A(s)X(s)\{ \int_0^s X^{-1}(\tau)F(\tau,y(\tau))du(\tau)\}ds$$

$$+ \int_0^t F(s,y(s))du(s). \tag{3.8}$$

Denote the first two integrals on the right hand side of (3.8) by I_1 I_2 respectively. By the definition of $X(t)$, we have

$$A(s)\, X(s) = \frac{d}{ds} X(s).$$

Therefore,

$$I_1 = \int_0^t A(s)X(s)x_0\, ds = \int_0^t dX(s)x_0 = X(t)x_0 - x_0. \tag{3.9}$$

Set

$$Y(t) = \int_0^t X^{-1}(s)\, F(s,y(s))du(s).$$

Then

$$I_2 = \int_0^t A(s)\, X(s)\, Y(s)ds = \int_0^t dX(s)Y(s). \tag{3.10}$$

Integrate by parts the last integral in (3.10) to obtain

$$I_2 = X(t)\, Y(t) - X(o)\, Y(o) - \int_0^t X(s)dY(s).$$

But $Y(o) = 0$, and by Lemma 1.1,

$$\int_o^t X(s)dY(s) = \int_o^t X(s)X^{-1}(s)\ F(s,y(s))du(s) = \int_o^t F(s,y(s))du(s).$$

Therefore,

$$I_2 = \int_o^t X(t)X^{-1}(s)\ F(s,y(s))du(s) - \int_o^t F(s,y(s))du(s) \quad (3.11)$$

From (3.8), (3.9) and (3.11), it follows that

$$x_o + \int_o^t A(s)y(s)ds + \int_o^t F(s,y(s))du(s)$$

$$= x_o + X(t)x_o - x_o + \int_o^t X(t)X^{-1}(s)\ F(s,y(s))du(s)$$

$$- \int_o^t F(s,y(s))du(s) + \int_o^t F(s,y(s))du(s)$$

$$= x(t) + \int_o^t X(t)X^{-1}(s)F(s,y(s))\ du(s) = y(t),$$

as desired.

q. e. d.

Theorem 3.2. Let S be a closed, bounded convex, non-empty subset of a Banach space X. Suppose that P and Q map S into X and that

(i) $Px + Qy \in S$ for all $x, y \in S$;

(ii) P is completely continuous ;

(iii) Q is a contraction.

Then, there exists $z \in S$ such that $Pz + Qz = z$.

Assume the following hypotheses :

(A_1) $X(t)X^{-1}(s)$ is integrable in s with respect to u for fixed $t \in J$ and

$$\sup_{t \geq 0} \sup_{\pi} \left[\sum_{i=1}^{N} \left\{ \int_{t_o}^{t_{i-1}} |X(t_i)-X(t_{i-1})|\ |X^{-1}(s)|dv_u(s) \right.\right.$$

$$+ \int_{t_{i-1}}^{t_i} |X(t_i)X^{-1}(s)|\, dv_u(s)\}\Big] \le a_o < \infty ,$$

where $\pi : 0 = t_o < t_1 < \dots < t_N = t$ denotes any partition of $[0,t]$, $t \in J$ and a_o is a positive constant.

(A_2) $z \to F(t,z)$ is a continuous map from $BV(J)$ to $BV(J)$ and $z_k \to z$ pointwise as $k \to \infty$ implies $F(t,z_k(t)) \to F(t,z(t))$ uniformly in $t \in J$.

Theorem 3.3. Let (A_1) and (A_2) hold. Assume that

(i) $F(t,z)$, $G(t,z)$ are integrable with respect to u and $F(t,0) \equiv 0 \equiv G(t,0)$;

(ii) for each $\gamma > 0$, there is a $\delta > 0$ such that

$$|F(t,z)| \le \gamma |z|$$

uniformly in $t \ge 0$, whenever $|z| \le \delta$;

(iii) for each $\xi > 0$, there is an $\eta > 0$ such that

$$|G(t,z_1) - G(t,z_2)| \le \xi |z_1 - z_2|$$

uniformly in $t \ge 0$, whenever $|z_1| \le \eta$, $|z_2| \le \eta$.

Suppose that $x \in BV(J)$ is the solution of (3.1). Then, there is a number ε_o with the following property, for each ε, $0 < \varepsilon \le \varepsilon_o$, there corresponds a δ_o such that whenever $\|x\|^* \le \delta_o$, there exists at least one solution $z(t)$ of (3.3) such that $z \in BV(J)$ and $\|z\|^* \le \varepsilon$.

Proof. Fix a $\xi > 0$ such that $\xi a_o < 1$ and choose an $\eta > 0$ such that $|G(t,z_1) - G(t,z_2)| \le \xi |z_1 - z_2|$, uniformly in $t \ge 0$ whenever $|z_1| \le \eta$, $|z_2| \le \eta$. Let $\gamma = (1-\xi a_o)/2a_o$. For this γ, select a $\delta > 0$ such that condition (ii) of the theorem holds. Let $\varepsilon_o = \min\{\eta, \delta\}$. For $\varepsilon \in (0, \varepsilon_o]$, define

$$S(\varepsilon), \{z \in BV(J) : \|z\|^* \le \varepsilon\}.$$

Then $S(\varepsilon)$ is a closed, convex subset of $BV(J)$. Let P and Q be the operators on $S(\varepsilon)$ given by

$$(Pz)(t) = x(t) + \int_0^t X(t)X^{-1}(s)F(s,z(s))du(s) \tag{3.12}$$

$$(Qz)(t) = \int_0^t X(t)X^{-1}(s)G(s,z(s))du(s). \tag{3.13}$$

For $z_1, z_2 \in S(\varepsilon)$, we have

$$\| Pz_1 + Qz_2 \|^* \leq \| x \|^* + \sup_{t \geq 0} \sup_{\pi} \Big[\sum_{i=1}^{N} \{ \int_0^{t_{i-1}} |X(t_i) - X(t_{i-1})| \, |X^{-1}(s)|$$

$$(\gamma |z_1(s)| + \xi |z_2(s)|) \, dv_u(s) + \int_{t_{i-1}}^{t_i} |X(t_i)X^{-1}(s)|$$

$$(\gamma |z_1(s)| + \xi |z_2(s)|) \, dv_u(s) \} \Big]$$

$$\leq \delta_0 + (\gamma + \xi) \, \varepsilon \, a_0$$

$$< \varepsilon ,$$

provided $\delta_0 < (1 - \xi a_0)(\varepsilon/2)$. Thus $Pz_1 + Qz_2 \in S(\varepsilon)$ for all $z_1, z_2 \in S(\varepsilon)$. Next,

$$\| Qz_1 - Qz_2 \|^* \leq \xi a_0 \| z_1 - z_2 \|^* ,$$

together with the fact that $\xi a_0 < 1$ implies that Q is a contraction on $S(\varepsilon)$. To see that P is completely continuous on $S(\varepsilon)$, let $\{z_k\}$ be any sequence in $S(\varepsilon)$. Then $\{z_k\}$ is uniformly bounded and is of uniform bounded variation. Therefore, by Theorem 1.1, there is a subsequence $\{z_{k_j}\}$ which converges point-wise to some $z^* \in BV(J)$. Hence,

$$\| Pz_{k_j} - Pz^* \|^* \leq \sup_{0 \leq s < \infty} |F(s, z_{k_j}(s)) - F(s, z^*(s))| a_0$$

$$\to 0 \text{ as } j \to \infty \text{ by } (A_2).$$

This means that $\{Pz_k\}$ is relatively compact and consequently P is completely continuous. The desired conclusion now follows from Theorem 3.2.

q. e. d.

Corollary 3.1. In addition to conditions of Theorem 3.3, if $\lim_{t \to \infty} x(t) = 0$ and if

$$(A_3) \qquad \lim_{t \to \infty} \int_0^T |X(t)X^{-1}(s)|dv_u(s) = 0 \quad \text{for each } T > 0$$

then $\lim_{t \to \infty} z(t) = 0.$

Proof. The set $S_0(\varepsilon) = \{z \in S(\varepsilon) : \lim_{t \to \infty} z(t) = 0\}$ is a closed subspace of $S(\varepsilon)$. Therefore, it suffices to prove that $Pz_1 + Qz_2 \in S_0(\varepsilon)$ whenever $z_1, z_2 \in S_0(\varepsilon)$. Let $\beta > 0$ be given. Choose $T_1 > 0$ so that

$$|x(t)| < \beta/5, \ |z(t)| < \beta/5 \quad \text{for } t \geq T_1.$$

Select $T_2 \geq T_1$ such that (by (A_3)),

$$\int_0^{T_1} |X(t)X^{-1}(s)|dv_u(s) < \beta\, a_0/5\varepsilon \ , \text{ for } t \geq T_2.$$

Then, for $z_1, z_2 \in S_0(\varepsilon)$ and $t \geq T_2$, we have

$$\begin{aligned}|(Pz_1 + Qz_2)(t)| \leq |x(t)| &+ \int_0^{T_1} |X(t)X^{-1}(s)| \ \gamma\, |z_1(s)|dv_u(s)\\ &+ \int_{T_1}^{t} |X(t)X^{-1}(s)| \ \gamma\, |z_1(s)|dv_u(s)\\ &+ \int_0^{T_1} |X(t)X^{-1}(s)| \, \xi\, |z_2(s)|dv_u(s)\\ &+ \int_{T_1}^{t} |X(t)X^{-1}(s)| \, \xi\, |z_2(s)|dv_u(s)\end{aligned}$$

$$\leq \beta/5 + \gamma\, \varepsilon \frac{\beta}{5\varepsilon} a_0 + \gamma\, \frac{\beta}{5} \int_0^t |X(t)X^{-1}(s)|dv_u(s)$$

$$+ \xi\, \varepsilon \frac{\beta}{5\varepsilon} a_0 + \xi \frac{\beta}{5} \int_0^t |X(t)X^{-1}(s)|dv_u(s)$$

$$< \frac{\beta}{5} + \frac{\beta}{5} \gamma a_o + \frac{\beta}{5} \gamma a_o + \frac{\beta}{5} \xi a_o + \frac{\beta}{5} \xi a_o < \beta ,$$

since $\xi a_o < 1$ by the choice of ξ in Theorem 3.3. It follows that $(Pz_1 + Qz_2)(t) \to 0$ and $t \to \infty$ and hence $Pz_1 + Qz_2 \in S_o(\varepsilon)$.

q. e. d.

A sufficient condition for $\| x \|^* \leq \delta_o$ is that $|x_o| \leq \delta_o(1 + b_o)^{-1}$, where b_o is a positive constant such that,

$$\sup_{t \geq 0} \sup_{\pi} \left[\sum_{i=1}^{N} \int_{t_{i-1}}^{t_i} |A(s)X(s)|ds \right] \leq b_o \tag{3.14}$$

Thus Theorem 3.3 may be regarded as a stability result for the system (3.3) in the following sense : for any $\varepsilon_1 > 0$, there exists $\varepsilon_2 > 0$ such that $|x_o| \leq \varepsilon_2$ implies that the solution z(t) of (3.3) $\in BV(J)$ and $\| z \|^* \leq \varepsilon_1$.

Corollary 3.1 is a type of asymptotic stability result. Hypothesis (A_1) clearly implies

$$\sup_{t \geq 0} V(\int_o^t |X(t)X^{-1}(s)|dv_u(s), [0,t]) \leq a_o ,$$

though the converse is not true in general.

Example 3.1. Let $A(t) = -(t+1)^{-1}$ be the scalar function. The solution $x(t) = x(t,o,x_o)$ of (3.1) is given by $x(t) = x_o(t+1)^{-1}$, which is clearly a function of bounded variation on $[0, \infty)$. Here $X(t)X^{-1}(s) = (t+1)^{-1}(s+1)$. Obviously, (3.14) holds. Choose

$$u(t) = \begin{cases} 0 & , \ 0 \leq t < 1 \\ (t+1)^{-2} & , \ 1 \leq t < \infty \end{cases} .$$

Then (A_1) and (A_3) are satisfied.

3.2 Aymptotic Equivalence

We now give criteria for asymptotic equivalence between solutions of (3.1) and (3.2) under the hypotheses (A_1) - (A_3) of Section 3.1 and various conditions on the nonlinearity F. At the end of the section, we also present a result on the asymptotic uniqueness of solutions of (3.5).

<u>Theorem 3.4.</u> Let $(A_1) - (A_3)$ hold. Suppose F satisfies

$$|F(t,y)| \leq m(t)|y|, \quad t \in J, \quad 0 \leq y < \infty$$

where $m(t)$ is bounded, dv_u- integrable function on J such that $m(t) \to 0$ as $t \to \infty$ and $M a_0 < 1/2$, $M = \sup_{t \in J} m(t)$.

Then the systems (3.1) and (3.2) are asymptotically equivalent in the following sense : to each bounded solution $x(t) \in BV(J)$ of (3.1), there corresponds a solution $y(t) \in BV(J)$ of (3.2) such that

$$\lim_{t \to \infty} |y(t) - x(t)| = 0 , \tag{3.15}$$

and conversely, to each bounded solution $y(t) \in BV(J)$ of (3.2), there corresponds a solution $x(t) \in BV(J)$ of (3.1) such that (3.15) holds.

<u>Proof.</u> Let $x(t) \in BV(J)$ be a bounded solution of (3.1) with $\| x \|^* \leq b$, $b > 0$ and let P be the operator defined by (3.12) acting on $S_b = \{y \in BV(J) : \| y \|^* \leq b\}$. Then $P(S_{2b}) \subseteq S_{2b}$, for if $\| y \|^* \leq 2b$, we have

$$\| Py \|^* \leq \| x \|^* + \sup_{t \geq 0} \sup_{\pi} \left[\sum_{i=1}^{N} \{ \int_{t_0}^{t_{i-1}} |X(t_i)-X(t_{i-1})| \, |X^{-1}(s)| m(s)| \, |y(s)| dv_u(s) \right.$$

$$\left. + \int_{t_{i-1}}^{t_i} |X(t_i)X^{-1}(s)| |m(s)| |y(s)| dv_u(s) \} \right]$$

$$\leq b + 2 b M a_0 \leq 2b ,$$

since $M a_0 \leq 1/2$. From our hypotheses, it is clear that P is continuous. Relative compactness of $P(S_{2b})$ follows by the method similar to the one used in establishing the complete continuity of P in Theorem 3.3. Therefore, by Schauder's fixed point theorem, there is a function $y(t)$ such that

$$y(t) = x(t) + \int_0^t X(t)X^{-1}(s)F(s,y(s))du(s) \tag{3.16}$$

and $\| y \|^* \leq 2b$. To see that (3.15) holds, let $\delta > 0$ be given. Since $m(t) \to 0$ as $t \to \infty$, there is a $T > 0$ such that

$$m(t) \leq \delta / 4 \ ba_o , \quad t \geq T.$$

By (A_3), select $T_1 \geq T$ such that

$$\int_o^T |X(t)X^{-1}(s)| \, dv_u(s) \leq \delta/4 \, bM, \quad t \geq T.$$

Then for $t \geq T_1$, we have

$$|y(t)-x(t)| \leq \int_o^T |X(t)X^{-1}(s)| \, m(s) \mid y(s) \mid dv_u(s)$$

$$+ \int_T^t |X(t)X^{-1}(s)| \, m(s) \mid y(s) \mid dv_u(s)$$

$$\leq 2\, bM \int_o^T |X(t)X^{-1}(s)| \, dv_u(s)$$

$$+ 2\, b \sup_{t \geq T} m(t) \int_o^t |X(t)X^{-1}(s)| \, dv_u(s)$$

$$\leq \delta/2 + \delta/2 = \delta .$$

Since $\delta > 0$ is arbitrary, (3.15) is established and therewith the first half of the theorem is proved. To prove the second half, let $y(t) \in BV(J)$ be a bounded solution of (3.2). Define

$$x(t) = y(t) - \int_o^t X(t)X^{-1}(s)F(s,y(s))du(s), \quad t \in J. \tag{3.17}$$

Since $\| x \|^* \leq \| y \|^* + M \, a_o \, \| y \|^*$, it is clear that $x(t) \in BV(J)$. Validity of (3.15) may be established as in the first half. It remains to show that $x(t)$ is a solution of (3.1). This is obvious in the case of ordinary differential equations (that is, when u is an absolutely continuous function), but not so in the present case. We prove that

$$x(t) = x_o + \int_o^t A(s) \, x(s) \, ds. \tag{3.18}$$

From (3.17), we have

$$x_0 + \int_0^t A(s)x(s)ds = x_0 + \int_0^t A(s)y(s)ds - I, \qquad (3.19)$$

where $\quad I = \int_0^t A(s)X(s) \{ \int_0^s X^{-1}(r)F(r,y(r))du(r) \} ds.$

As in the proof of Theorem 3.1, it can be shown that

$$I = \int_0^t X(t)X^{-1}(s)F(s,y(s))du(s) - \int_0^t F(s,y(s))du(s). \qquad (3.20)$$

Substituting (3.20), into the right hand side of (3.19), it follows from Theorem 2.1 that (3.18) reduces to the identity $x(t) = x(t)$, which indeed, was our objective.

q. e. d.

Theorem 3.5. Let (A_1) and (A_2) hold. Suppose that

(i) there exist positive constants K and a such that

$$|X(t)X^{-1}(s)| \le K e^{-a(t-s)}, \quad t \ge s \ge 0; \qquad (3.21)$$

(ii) there exist a nonnegative dv_u-integrable function $h(t)$ and a positive constant α such that

$$|F(t,y)| \le h(t) \quad \text{for} \quad |y| \le \alpha, \quad t \in J$$

and $\|x\|^* + h_0 \le \alpha$, where h_0 is a positive constant satisfying

$$\sup_{t \ge 0} \sup_{\pi} \left[\sum_{i=1}^{N} \left\{ \int_0^{t_{i-1}} |X(t_i)-X(t_{i-1})| \, |X^{-1}(s)| \, h(s)dv_u(s) + \int_{t_{i-1}}^{t_i} |X(t_i)X^{-1}(s)| \, h(s)dv_u(s) \right\} \right] \le h_0 \qquad (3.22)$$

Then the systems (3.1) and (3.2) are asymptotically equivalent in the sense of Theorem 3.4.

Proof. We only establish the validity of (3.15), the rest of the proof being similar to that of Theorem 3.4. From (3.21) and (3.22), we note that

$$\int_0^\infty h(s)\, dv_u(s) < \infty \tag{3.23}$$

whence

$$\lim_{t \to \infty} \int_t^\infty h(s)\, dv_u(s) = 0. \tag{3.24}$$

Therefore,

$$|y(t)-x(t)| \le K\, e^{-at} \int_0^{t/2} e^{as}\, h(s)dv_u(s) + K\, e^{-at} \int_{t/2}^t e^{as} h(s)dv_u(s)$$

$$\le K\, e^{-at/2} \int_0^{t/2} h(s)dv_u(s) + K \int_{t/2}^t h(s)dv_u(s). \tag{3.25}$$

By (3.25), the first integral on the right hand side of (3.25) approaches a finite limit as $t \to \infty$ and, by (3.24), the second integral tends to zero as $t \to \infty$. Therefore $|y(t)-x(t)| \to 0$ as $t \to \infty$.

q. e. d.

Theorem 3.6. The conclusion of Theorem 3.5 remains valid if the conditions (i) and (ii) are replaced by

(i)' there exists a positive constant K such that

$$|X(t)X^{-1}(s)| \le K,\ t \ge s \ge 0 \quad \text{and} \quad \lim_{t \to \infty} |X(t)| = 0 ;$$

(ii)' F satisfies

$$|F(t,y)| \le \psi\,(t, |y|\,) ,$$

where $\psi(t,r)$ is a dv_u-integrable function for $0 \le t,\ r < \infty$, which is monotone non-decreasing in r for each fixed t and is such that

$$\sup_{t\ge 0}\ \sup_{\pi} \Bigg[\sum_{i=1}^N \Big\{ \int_0^{t_{i-1}} |X(t_i)-X(t_{i-1})|\,|X^{-1}(s)|\ \psi(s,c)dv_u(s)$$

$$+ \int_{t_{i-1}}^{t_i} |X(t_i)X^{-1}(s)|\ \psi(s,c)dv_u(s)\ \Big\} \Bigg] < \infty,$$

for each $c > 0$.

We now consider the asymptotic uniqueness of solution of (3.5).

Theorem 3.7. Suppose that

(i) u is of type φ ;

(ii) for $(t,x,y) \in J \times R^n \times R^n$ and for $h > 0$, there is an $\alpha > 0$ such that

$$| x-y + h[f(t,x) - f(t,y)] | \leq (1-\alpha h)|x-y| ;$$

(iii) $|g(t,y)| \leq \omega(t,|y|)$ for $(t,y) \in J \times R^n$, where $\omega : J \times J \to J$, $\omega(r,s) \leq \omega(r,t)$ for $s \leq t$ and $\int_0^\infty \omega(s,c)\, dv_u(s) < \infty$ for each $c > 0$.

If x and y are bounded solutions of (3.4) and (3.5) respectively on J such that $x(o) = y(o)$, then

$$\lim_{t \to \infty} |y(t) - x(t)| = 0.$$

Proof. Let M and N be positive numbers such that $|x(t)| \leq M$, $|y(t)| \leq N$ for $t \in J$. Set

$$m(t) = |y(t)| - |x(t)|,\ t \in J.$$

Then $m(o) = 0$ and for $t \in [t_{k-1}, t_k)$, we have

$$m'(t) = \lim_{s \to 0_+} \frac{m(t+s) - m(t)}{s}$$

which in view of (ii) and (iii) yields

$$m'(t) \leq -\alpha\, m(t) + \omega(t,N)\, |u'(t)|.$$

This implies

$$m(t) \leq m(t_{k-1})e^{-\alpha(t-t_{k-1})} + e^{-\alpha t} \int_{[t_{k-1},t]} e^{\alpha s}\omega(s,N)|u'(s)|ds,\ t \in [t_{k-1},t_k]. \tag{3.26}$$

At the points t_k of discontinuity $k = 1,2,\ldots$, we have

$$m(t_k) \leq |m(t_k) - m(t_k-)| + \lim_{s \to 0_+} m(t_k-s).$$

Using (3.26) and the fact that $x(t_k) = x(t_k-)$, we obtain after a little computation,

$$m(t_k) \leq m(t_{k-1})\, e^{-\alpha(t_k - t_{k-1})} + \omega(t_k, N)\, |a_k| + e^{-\alpha t_k} \int_{[t_{k-1}, t_k)} e^{\alpha s}\, \omega(s,N)|u'(s)|ds, \qquad (3.27)$$

where $a_k = u(t_k) - u(t_k-)$.

Now, inequality (3.26) gives (note that $m(t_o) = m(o) = 0$)

$$m(t) \leq e^{-\alpha t} \int_{[0,t]} e^{\alpha s}\, \omega(s,N)|u'(s)|ds, \quad \text{for} \quad t \in [0,t_1)$$

and

$$m(t) \leq m(t_1)\, e^{-\alpha(t-t_1)} + e^{-\alpha t} \int_{[t_1,t]} e^{\alpha s}\, \omega(s,N)|u'(s)|ds,$$

$$\text{for } t \in [t_1, t_2).$$

Hence, from (3.27), we have

$$m(t) \leq \omega(t_1,N)|a_1| + \sum_{i=1}^{2} e^{-\alpha t} \int_{[t_{i-1}, t_i)} e^{\alpha s}\omega(s,N)|u'(s)|ds,$$

$$\text{for } t \in [\ 0,t_2\ )$$

and in general, for $t \geq 0$, where $0 < t_1 < t_2 < \ldots < t_k = t$,

$$m(t) \leq \sum_{i=1}^{k-1} \omega(t_i,N)|a_i| + \sum_{i=1}^{k} e^{-\alpha t} \int_{[t_{i-1}, t_i)} e^{\alpha s}\, \omega(s,N)|u'(s)|ds$$

$$= e^{-\alpha t} \int_0^t e^{\alpha s}\omega(s,N)dv_u(s)$$

$$\to 0 \quad \text{as } t \to \infty \quad \text{as in the proof of Theorem 3.5.}$$

q. e. d.

Corollary 3.2. If y_1 and y_2 are any two bounded solutions of (3.5) such that $y_1(o) = y_2(o)$, then under the conditions of Theorem 3.7, $|y_1(t) - y_2(t)| \to 0$ as $t \to \infty$.

Examples below illustrate Theorems 3.4 - 3.7.

Example 3.2. Consider the system (3.1), where

$$x = \begin{bmatrix} x_1 \\ x_2 \end{bmatrix} \quad \text{and} \quad A(t) = \begin{bmatrix} -1 & e^t \\ -e^t & -1 \end{bmatrix}.$$

A fundamental matrix of (3.1) satisfies

$$X(t) = e^{-t} \begin{bmatrix} \sin e^t & -\cos e^t \\ \cos e^t & \sin e^t \end{bmatrix}, \quad X^{-1}(t) = e^t \begin{bmatrix} \sin e^t & \cos e^t \\ -\cos e^t & \sin e^t \end{bmatrix}.$$

$$\text{Let } F(t,y) = \begin{bmatrix} e^{-t}y_1 - (t+1)^{-1} e^{-2t} y_2 \\ (2t+3)^{-1} e^{-2t} y_1 + e^{-2t} y_2 \end{bmatrix}, \quad y = \begin{bmatrix} y_1 \\ y_2 \end{bmatrix}.$$

Since $|F(t,y)| \leq 2e^{-t} |y|$, we may take $m(t) = 2e^{-t}$ in Theorem 3.4. Choose

$$u(t) = \begin{cases} 0 & , \ 0 \leq t < 1 \\ \beta e^{-t} & , \ 1 \leq t < \infty \end{cases}, \quad \beta > 0.$$

It is possible to have $Ma_0 \leq 1/2$ if β is suitably chosen (indeed, it suffices to take $0 < \beta < e/32$).

Example 3.3. Let $A(t) = -a$, where a is a positive real number. Then $X(t)\, X^{-1}(s) = e^{-a(t-s)}$, $t \geq s \geq 0$. Take $F(t,y) = e^{-t} \sin (e^{ty^2}) - t^2 e^t$, $0 \leq t, y < \infty$. Since $|F(t,y)| \leq 1 + t^2 e^t$ for all t and y, we may take $h(t) = 1 + t^2 e^t$ in Theorem 3.5. Choose

$$u(t) = \begin{cases} 0 & , \ 0 \leq t < 1 \\ e^{-2t} & , \ 1 \leq t < \infty \end{cases}.$$

Then all the conditions of Theorem 3.5 are satisfied.

Example 3.4. Let $A(t)$ be as in Example 3.3 and $F(t,y) = t^3 e^{-2t} (y+1) + t(t+1)^{-1} e^{-t} + (t+2) e^{-3t}$. Then

$|F(t,y)| \leq \psi\ (t,|y|)$, where $\psi(t,r) = (t^3 r + t^3 + 2t + 2)e^{-t}$. All the conditions of Theorem 3.6 are satisfied if we choose

$$u(t) = \begin{cases} 0 & , \ 0 \leq t < 1 \\ t^2 & , \ 1 \leq t < \infty \end{cases}.$$

<u>Example 3.5.</u> Consider the equations

$$x' = -x + 1, \quad x(o) = 2 \tag{3.28}$$

and

$$Dy = -y + 1 + (1-y + e^{-2t})\ e^{-t}\ Du,\ y(o) = 2 \tag{3.29}$$

where

$$u(t) = \begin{cases} 0 & , \ 0 \leq t < 1 \\ t & , \ 1 \leq t < \infty \end{cases}.$$

The hypotheses of Theorem 3.7 are clearly satisfied. The solutions of (3.28) and (3.29) are respectively given by

$$x(t) = 1 + e^{-t}$$

and

$$y(t) = \begin{cases} 1 + e^{-t} & , \ 0 \leq t < 1 \\ 1 + e^{-t} + e^{-2t} & , \ 1 \leq t < \infty \end{cases}.$$

Notice that neither $x(t)$ nor $y(t)$ approach zero as $t \to \infty$, although $|x(t) - y(t)|$ does.

3.3 Stability of Weakly Nonlinear Measure Differential Systems

Consider the system

$$Dy = A(t)y + F(t,y) + G(t,y)Du \tag{3.30}$$

where $A(t)$ is continuous on J and F,G are continuous on $J \times R^n$, Let $y(t) = y(t,t_0,x_0)$ be a solution of (3.30) through (t_0,x_0), existing on the right of $t_0 \geq 0$ in $S(r) = \{x \in R^n : |x| \leq r\}$. In this section, we present results on uniform asymptotic stability of the null solution of (3.30).

<u>Definition 3.1.</u> The null solution of (3.30) is said to be eventually uniformly asymptotically stable if the following two

conditions are satisfied:

(i) for every $\varepsilon > 0$, there exist a $\delta = \delta(\varepsilon) > 0$ and a $\tau = \tau(\varepsilon) > 0$ such that $|y(t,t_0,x_0)| < \varepsilon$, $t \geq t_0 \geq \tau(\varepsilon)$, provided that $|x_0| \leq \delta$;

(ii) for every $\eta > 0$ there exist positive numbers δ_0, τ_0 and $T = T(\eta)$ such that $|y(t,t_0,x_0)| < \eta$, $t \geq t_0 + T$, $t_0 \geq \tau_0$, provided $|x_0| \leq \delta_0$.

We need the following two lemmas:

Lemma 3.1. Let m and K be scalar, nonnegative functions defined and integrable on [0,T] and let h be a scalar, nonnegative function which is of bounded variation on [0,T]. Then the inequality

$$m(t) \leq h(t) + \int_0^t K(s)m(s)ds, \quad t \in [0,T]$$

implies

$$m(t) \leq h(0)\exp\Big[\int_0^t K(s)ds\Big] + \int_0^t \exp\Big(\int_s^t K(r)dr\Big)dh(s), \quad t\in[0,T]$$

Lemma 3.2. Let m and K and g be nonnegative integrable functions on [0,T] and c be a nonnegative constant. Suppose that g(r) is monotonically increasing in r and g(o) = 0. Then, if the inequality

$$m(t) \leq c + \int_0^t K(s)\, g(m(s))ds, \quad t \in [0,T]$$

holds, the inequality

$$m(t) \leq \psi^{-1}\Big[\psi(c) + \int_0^t K(s)ds\Big], \quad t \in [0,T]$$

remains valid as long as $\psi(c) + \int_0^t K(s)ds$ lies in the domain of definition of ψ^{-1}, where ψ is defined by

$$\psi(\eta) = \int_\varepsilon^\eta \frac{d\tau}{g(\tau)}, \quad \varepsilon > 0, \quad \eta > 0,$$

and ψ^{-1} is the inverse mapping of the function ψ.

Theorem 3.8. Suppose that

(i) in (3.30), A(t) = A is a constant n by n matrix with all its characteristic roots having negative real parts;

(ii) given any $\varepsilon > 0$, there exist $\delta(\varepsilon), T(\varepsilon) > 0$ such that $|F(t,y)| \leq \varepsilon|y|$, whenever $t \geq T(\varepsilon)$ and $|y| < \delta(\varepsilon)$;

(iii) there exists $r > 0$ such that if $|y| \leq r$, then $|G(t,y)| \leq \gamma(t)$ for all $t \geq 0$, where $\gamma(t)$ is a continuous function satisfying

$$\Gamma(t) = \int_t^{t+1} \gamma(s)\, dv_u(s) \to 0 \quad \text{as } t \to \infty.$$

Then, there exist $T_o \geq 0$ and $\delta_o > 0$ such that for every $t_o \geq T_o$ and $|x_o| \leq \delta_o$, any solution $y(t) = y(t,t_o,x_o)$ of (3.30) satisfies $|y(t)| \to 0$ as $t \to \infty$. In particular, if (3.30) possesses the null solution, then it is eventually uniformly asymptotically stable.

Proof. Since all the characteristic roots of A have negative real parts, there exist positive constants c_1 and a such that the fundamental matrix $X(t) = e^{At}$ of $x' = Ax$ satisfies

$$|X(t)| \leq c_1\, e^{-at} \quad \text{for } t \geq 0.$$

Let $\delta = \delta((a/2)c_1^{-1}) < r$. Choose t_o and x_o so that $t_o \geq T = T((a/2)c_1^{-1})$ and $|x_o| < \delta_o < \delta$. Then, for $t \geq t_o$, we have from Theorem 3.1,

$$|y(t)| \leq c_1\, e^{-a(t-t_o)}\, |x_o| + \int_{t_o}^{t} (a/2)e^{-a(t-s)}\, |y(s)|ds$$

$$+ \int_{t_o}^{t} c_1\, e^{-a(t-s)}\, \gamma(s)\, dv_u(s),$$

as long as $|y(t)| \leq \delta$. An application of Lemma 3.1 yields

$$|y(t)| \leq c_1 e^{-(a/2)(t-t_o)}\, |x_o| + c_1 \int_{t_o}^{t} e^{-(a/2)(t-s)}\, \gamma(s)dv_u(s). \tag{3.31}$$

Suppose that $t_o \geq 1$. Then for $t \geq t_o$, by changing the order of integration, we obtain

$$\int_{t_o}^{t} e^{(a/2)s}\, \gamma(s)\, dv_u(s) \leq \int_{t_o-1}^{t} e^{(a/2)(s+1)}\, \Gamma(s)ds. \tag{3.32}$$

Set

$$Q(t) = \sup \{ \Gamma(s) : t - 1 \leq s < \infty \}.$$

Then $Q(t)$ decreases to zero as $t \to \infty$ and we have from (3.32),

$$\int_{t_0}^{t} e^{-(a/2)(t-s)} \gamma(s) dv_u(s) \leq e^{a/2} Q(t_0) \int_{t_0-1}^{t} e^{-(a/2)(t-s)} ds$$

$$\leq (2/a) e^{a/2} Q(t_0).$$

Thus, if δ_0 and T_0 are so chosen that

$$\delta_0 < \min \{ \delta, (\delta/2) c_1^{-1} \},$$

$$T_0 \geq \max \{1, T\},\ Q(t_0) \leq (\delta/2)(c_1(2/a) e^{a/2})^{-1},$$

then for $t \geq T_0, |x_0| \leq \delta_0$, we have $|y(t)| < \delta$, which implies the existence of $y(t,t_0,x_0)$ on the interval $[t_0, \infty)$. Finally, for $t \geq T_0 \geq 1$,

$$\int_{t_0}^{t} e^{-(a/2)(t-s)} \gamma(s)\, dv_u(s)$$

$$\leq e^{a/2} \int_{t_0-1}^{t} e^{-(a/2)(t-s)} \Gamma(s) ds$$

$$\leq e^{a/2} \{ \int_{0}^{t/2} e^{-(a/2)(t-s)} \Gamma(s) ds + \int_{t/2}^{t} e^{-(a/2)(t-s)} \Gamma(s) ds \}$$

$$\leq e^{a/2} \{ Q(1) \int_{0}^{t/2} e^{-(a/2)(t-s)} ds + Q((t/2)+1) \int_{t/2}^{t} e^{-(a/2)(t-s)} ds \}$$

$$\leq e^{a/2} \{ Q(1)(2/a) e^{-(a/4)t} + Q((t/2)+1)(2/a) \}$$

$$\to 0 \text{ as } t \to \infty.$$

Hence, from (3.31), $|y(t)| \to 0$ as $t \to \infty$. q. e. d.

Theorem 3.9. Suppose that all the conditions of Theorem 3.8 are satisfied except the condition (ii) which is replaced by

$$|F(t,y)| \leq K(t)\, g(|y|),$$

for $|y| \le r$, where $g(o) = 0$, $g(\tau)$ is monotonically increasing in τ and $K(t)$ is a nonnegative function such that $\int_0^\infty K(s)ds$ is finite.

If $\int_0^1 [dt/g(t)]$ is divergent and (3.30) possesses null solution, then it is eventually uniformly asymptotically stable.

<u>Proof.</u> We have

$$|y(t)| \le c_1|x_0| + c_1 \int_{t_0}^{t} K(s)g(|y(s)|)ds + c_1 \int_{t_0}^{t} e^{-a(t-s)} \gamma(s)dv_u(s) \tag{3.33}$$

as long as $|y(t)| \le r$. If $t \ge t_0 \ge 1$, we obtain as in the proof of Theorem 3.8,

$$\int_{t_0}^{t} c_1 e^{-a(t-s)} \gamma(s)dv_u(s) \le (c_1/a)Q(t_0)e^a.$$

Therefore, from (3.33), we have for $t \ge t_0 \ge 1$,

$$|y(t)| \le c_1|x_0| + c_2\, Q(t_0) + \int_{t_0}^{t} c_1\, K(s)g(|y(s)|)ds\ ,$$

where $c_2 = (c_1/a)e^a$. By Lemma 3.2, this yields

$$|y(t)| \le \psi^{-1}[\ \psi(c_1|x_0| + c_2\, Q(t_0)) + c_1 \int_{t_0}^{t} K(s)ds]. \tag{3.34}$$

The right hand side of (3.34) can be made arbitrarily small for all $t \ge t_0$ by choosing $|x_0|$ and $Q(t_0)$ small enough so that $\psi\,[c_1|x_0| + c_2Q(t_0)]$ is as small as we desire (it approaches $-\infty$ arbitrarily). This proves the eventual uniform stability of the null solution of (3.30).

To see that $|y(t)| \to 0$ as $t \to \infty$, whenever $|x_0|$ is sufficiently small, we first note, by eventual uniform stability, that there is a constant M such that, for t_0 sufficiently large,

$$|y(t)| < M \quad \text{for all} \quad t \ge t_0$$

whenever $|x_0| < \delta$, where M depends on δ and may be supposed to satisfy $M \le r$. Therefore,

$$|y(t)| \le c_1|x_0|e^{-a(t-t_0)} + c_1\, g(M)\int_0^{t/2} e^{-a(t-s)}K(s)ds$$

$$+ c_1 g(M) \int_{t/2}^{t} e^{-a(t-s)}\, K(s)ds + c_1 \int_{t_0}^{t} e^{-a(t-s)}\, \gamma\,(s)dv_u(s)$$

$$\le c_1|x_0|e^{-a(t-t_0)} + c_1 g(M)e^{-at/2} \int_0^{t/2} K(s)ds$$

$$+ c_1 g(M) \int_{t/2}^{t} K(s)ds + c_1 e^a \int_{t_0-1}^{t} e^{-a(t-s)}\, \Gamma(s)ds.$$

As in the proof of Theorem 3.8,

$$\int_{t_0-1}^{t} e^{-a(t-s)}\, \Gamma(s)ds \to 0 \quad \text{as } t \to \infty$$

Therefore, since $\int_0^\infty K(s)ds$ is finite, it follows that $|y(t)| \to 0$ as $t \to \infty$.

q. e. d.

The condition $|G(t,y)| \le \gamma\,(t)$ in Theorems (3.8) and (3.9) does not explicitly bring into play the essential role of the discontinuities of u. The following result, in the proof of which we use Corollary 2.1, gives a somewhat better picture. This obviously requires u to be of type $\mathcal{Y}$. Solution y(t) of (3.30) on J satisfying $y(t_0) = x_0$ is given by (Theorem 3.1),

$$y(t) = X(t,t_0)x_0 + \int_{t_0}^{t} X(t,s)F(s,y(s))ds + \int_{t_0}^{t} X(t,s)G(s,y(s))du(s) \quad (3.35)$$

where $X(t,s) = X(t)\, X^{-1}(s)$, $X(t,t) = E$.

<u>Theorem 3.10.</u> Suppose that

(i) $X(t,s) \le \exp\left(\int_s^t a(r)dr\right)$, $t \ge s \ge t_0$

where the function a is continuous on J;

(ii) there exist non-negative, continuous functions p and q on J such that

$$|F(t,y)| \le p(t)|y|, \quad |G(t,y)| \le q(t)|y|$$

for $t \ge t_0$ and $|y| \le r$, $r > 0$;

(iii) u is of type φ ; $|b_k| q(t_k) < 1$ for $k = 1,2,\ldots$ where $b_k = v_u(t_k) - v_u(t_k-)$ and the series $\sum_{k=1}^{\infty} b_k q(t_k)$ converges absolutely.

Then the following estimate exists for the solutions of (3.30):

$$|y(t)| \leq P^{-1}|x_0| \exp\left(\int_{t_0}^{t} [a(s)+p(s)+q(s)v_u'(s)]ds\right), \ t \in J \qquad (3.36)$$

where $P = \prod_{k=1}^{\infty} (1 - |b_k| \, q(t_k))$.

In particular, if (3.30) possesses the null solution and if

$$p, \ qv_u' \in L'(J) \quad \text{and} \quad \int_J a(s)ds = -\infty$$

then the null solution is uniformly asymptotically stable.

<u>Proof.</u> Setting

$$z(t) = |y(t)| \exp\left(- \int_{t_0}^{t} a(s)ds\right)$$

and using the conditions of the theorem, we have from (3.35),

$$z(t) \leq |x_0| + \int_{t_0}^{t} p(s)z(s)ds + \int_{t_0}^{t} q(s)z(s)dv_u(s).$$

An application of Corollary 2.1 yields (3.36), from which the desired conclusion follows.

q. e. d.

<u>Example 3.6.</u> Consider the scalar equation

$$Dy = - ty + e^{-t}y + (t^3+ 1)^{-1}y \ Du, \ t \in [0, \infty)$$

where

$$u(t) = t^2 + \sum_{k=1}^{\infty} \left(\sum_{i=1}^{k} (i-1)^2 \chi_{[k-1,k)}(t)\right).$$

Then $b_k = -(2k-1)$, $k = 1,2,\ldots;$ $\qquad X(t,s) = \exp(-2^{-1}(t^2-s^2))$.

All the conditions of Theorem 3.10 are satisfied if we choose $a(t) = -t$, $p(t) = e^{-t}$ and $q(t) = (t^3 + 1)^{-1}$.

3.4 Notes.

Theorem 3.1 is a variation of parameters formula for measure differential equations and is taken from Schmaedeke [45]. Theorem 3.2 is Krasnoselskii's fixed point theorem [18]. Theorem 3.3, Corollary 3.1 and the contents of Section 3.2 (except Theorem 3.7 and Corollary 3.2, which are due to V.S.H. Rao [41]) are taken from the work of Pandit and Deo [34]. Lemma 3.1 is due to Barbashin [1], in which more results on the stability with respect to impulsive perturbations are also given. See also Zabalischin [52] and Pavlidis [35] for additional results. Lemma 3.2, which is a nonlinear generalization of Gronwall's inequality is due to Bihari and is given in Lakshmikantham and Leela [19]. Theorems 3.8 and 3.9 are taken from Raghavendra and M.R.M. Rao [38], which also contains stability results for the general nonlinear system (3.5). Theorem 3.10 is from Pandit [29].

CHAPTER 4

IMPULSIVE SYSTEMS

In the previous chapters, the equation

$$Dx = F(t,x) + G(t,x)Du \qquad (4.1)$$

was studied mainly by treating it as an impulsive perturbation of an <u>ordinary</u> system. The study of (4.1) becomes quite interesting if it is viewed as a perturbation of a system which itself contains impulses. In this chapter we consider the equation

$$Dy = A(t)y\,Du + f(t,y) + g(t,y)Du, \quad y(t_o) = x_o \qquad (4.2)$$

as a perturbed system of the linear measure differential equation

$$Dx = A(t)x\,Du, \quad x(t_o) = x_o \qquad (4.3)$$

where x,y are n-vectors, A(t) is an n by n matrix, continuous on $J = [t_o, \infty)$, $t_o \geq 0$; u is a right-continuous function of bounded variation on every compact subinterval of J; $f : J \times R^n \to R^n$ is Lebesgue integrable and $g : J \times R^n \to R^n$ is integrable with respect to the Lebesgue - Stieltjes measure du.

To simplify the analysis, throughout this chapter (except in the last section, which contains some general results), we shall assume that u is of type $\mathcal{Y}$. More specifically, u is assumed to be of the form

$$u(t) = t + \sum_{k=1}^{\infty} a_k H_k(t), \quad H_k(t) = \begin{cases} 0 & \text{if } t < t_k \\ 1 & \text{if } t \geq t_k \end{cases} \qquad (4.4)$$

where a_k are real numbers, Generally, a right-continuous function of bounded variation contains an absolutely continuous part and a singular part. The latter usually resembles

$$\sum_{k=1}^{\infty} a_k H_k(t)$$

when the discontinuities (which are obviously countable) are isolated. The above assumption is thus reasonable. Moreover, in this case, the predominqnt effect of the impulses is explicitly seen. It follows that

$$Du = 1 + \sum_{k=1}^{\infty} a_k \delta(t_k)$$

where $\delta(t_k)$ is the Dirac measure concentrated at t_k (note that $u' \equiv 1$ almost everywhere on J). For any $t \in J$, there is a unique integer $k \geq 1$ such that $t \in [t_{k-1}, t_k)$.

4.1 The Linear System.

The investigation of properties of solutions of (4.2) requires the preknowledge of solutions of the linear system (4.3). To this end, let $X(t)$ be a fundamental solution matrix of (4.3) and $X(t,s) = X(t)X^{-1}(s)$, $t, s \in J$ denote the transition matrix associated with $A(t)$. Then $X(t,t) = E$, $t \in J$. Moreover,

$$X(t,s)\, X(s,r) = X(t,r), \quad t, s, r \in J \tag{4.5}$$

and, in particular,

$$X(t,s) = X^{-1}(s,t), \quad t, s \in J. \tag{4.6}$$

Let $B_k = B_k(t_k)$ denote the matrix $E - a_k A(t_k)$, $k = 1,2,\ldots$

<u>Theorem 4.1.</u> Let B_k be non-singular for each k. Then for $t \in [t_{k-1}, t_k)$ and for any $x_0 \in R^n$, the (unique) solution $x(t) = x(t,t_0,x_0)$ of (4.3) is given by

$$x(t) = X(t,t_{k-1}) \left[\prod_{i=1}^{k-1} B_{k-i}^{-1}\, X(t_{k-i}, t_{k-i-1}) \right] x_0 . \tag{4.7}$$

(Here, the product $\prod_{i=1}^{k-1}$ is to be understood as E if $k = 1$).

<u>Proof.</u> Firstly, let $t \in [t_0, t_1)$. Then $u(t) = t$ and we have

$$x(t) = X(t,t_0)x_0. \tag{4.8}$$

At $t = t_1$, we know that

$$x(t_1) = x_0 + \int_{t_0}^{t_1} A(s)x(s)\, du(s)$$

$$= x(t_1-h) + \int_{t_1-h}^{t_1} A(s)x(s)du(s),$$

where $h > 0$ is sufficiently small. Letting $h \to 0_+$ and using the fact that $X(.,s)$ is continuous, we obtain

$$x(t_1) = \lim_{h\to 0+} [x(t_1-h) + \int_{t_1-h}^{t_1} A(s)x(s)du(s)]$$

$$= X(t_1,t_0)\, x_0 + a_1\, A(t_1)x(t_1)\ ,$$

which, since B_1 is invertible, yields

$$x(t_1) = B_1^{-1}\, X(t_1,t_0)x_0\ . \tag{4.9}$$

For $t \in [t_1,t_2)$, we note that

$$x(t) = X(t,\ t_1)x(t_1)\ , \tag{4.10}$$

where $x(t_1)$ is determined by (4.9). Using an argument similar to the earlier one, it follows that

$$x(t_2) = X(t_2,t_1)\ x(t_1) + a_2\, A(t_2)x(t_2),$$

whence

$$x(t_2) = B_2^{-1}\, X(t_2,t_1)\ B_1^{-1}\, X(t_1,t_0)x_0\ ,$$

which is (4.7) for $t = t_2$. The general case follows by induction.

q. e. d.

<u>Remark 4.1.</u> If $a_k = 0$ for each k (in which case $B_k = E$ is clearly invertible), then from (4.5), it is clear that (4.7) reduces to $x(t) = X(t,t_0)x_0$, which obviously solves the classical system $x' = A(t)x$, $x(t_0) = x_0$. If $a_k \neq 0$, then a sufficient condition for B_k to be invertible is that a_k^{-1} is not an eigenvalue

of $A(t_k)$.

Remark 4.2. Suppose that $A(t) = A$ is a constant n by n matrix. Then $X(t,s) = e^{(t-s)A}$. Moreover, $e^{(t_k-t_{k-1})A}$ and B_k^{-1} commute with each other for all k, since their inverses, namely, $e^{(t_{k-1}-t_k)A}$ and B_k respectively do so. This latter fact may be easily verified by expanding $e^{(t_{k-1}-t_k)A}$ in Taylor series. In this case, (4.7) assumes the form

$$x(t) = \left[\prod_{i=1}^{k-1} B_{k-i}^{-1} \right] e^{(t-t_0)A} x_0 \ . \tag{4.11}$$

The following theorem shows that, at the points t_k of discontinuity of u, solutions of (4.3) may exhibit an abrupt behaviour (see also Example 4.1).

Theorem 4.2. Let B_i, $i = 1,2,\ldots, j-1$ be non-singular for some $j > 1$, and let a_j^{-1} be an eigenvalue of $A(t_j)$. Then

(i) if x_0 is arbitrary, $x(t)$ does not exist at $t = t_j$;

(ii) if $x_0 = 0$, $x(t)$ is arbitrarily determined at $t = t_j$, meaning thereby that there are infinitely many solutions of (4.3) at $t = t_j$.

Proof. Since B_i is non-singular for $i = 1,2,\ldots, j-1$, $x(t_{j-1})$ exists by Theorem 4.1, and is given by

$$x(t_{j-1}) = \left[\prod_{i=1}^{j-1} B_{j-i}^{-1} X(t_{j-i}, t_{j-i-1}) \right] x_0 .$$

Further,

$$x(t_j) = x(t_j, t_{j-1}) x(t_{j-1}) + a_j A(t_j) x(t_j) ,$$

that is,

$$B_j x(t_j) = X(t_j, t_{j-1}) \left[\prod_{i=1}^{j-1} B_{j-i}^{-1} X(t_{j-1}, t_{j-i-1}) \right] x_0 \tag{4.12}$$

Therefore, if x_0 is such that the right hand side of (4.12) does not vanish, then $x(t_j)$ does not exist, since a_j^{-1} is an eigenvalue of $A(t_j)$. On the other hand, if $x_0 = 0$, then (4.12) implies

$$B_j x(t_j) = 0. \tag{4.13}$$

Since a_j^{-1} is an eigenvalue of $A(t_j)$, a non-trivial $x(t_j)$ satisfies (4.13), and, $cx(t_j)$ also satisfies it where c is any arbitrary constant.

q. e. d.

<u>Example 4.1.</u> Let $J = [0, \infty)$. Consider the system (4.3) where

$$A(t) = A = \begin{bmatrix} 2 & 1 \\ 3 & 4 \end{bmatrix}.$$

In (4.4), take $t_k = k$, $k = 1,2,\ldots,$; $a_1 = 2$, $a_2 = 1$ and $a_k = k^2$ for $k \geq 3$. Then a_2^{-1} is an eigenvalue of A, but a_1^{-1} is not. Choose $x_0 = \begin{bmatrix} 1 \\ 3 \end{bmatrix}$. Theorem 4.1 asserts the existence of a unique solution of (4.3) through $(0, x_0)$ on the interval $[0,2)$. In fact, it is not hard to see that the solution is given by

$$x(t) = \begin{bmatrix} x_1(t) \\ x_2(t) \end{bmatrix},$$

where

$$x_1(t) = \begin{cases} e^{5t}, & \text{for } 0 \leq t < 1 \\ -\dfrac{e^{5t}}{9}, & \text{for } 1 \leq t < 2 \end{cases}, \quad x_2(t) = \begin{cases} 3\, e^{5t}, & \text{for } 0 \leq t < 1 \\ -\dfrac{e^{5t}}{3}, & \text{for } 1 \leq t < 2 \end{cases}.$$

But $x(t)$ does not exist at $t = 2$. However, if we take $x_0 = \begin{bmatrix} 0 \\ 0 \end{bmatrix}$, then $x(t) = \begin{bmatrix} 0 \\ 0 \end{bmatrix}$ for $0 \leq t < 2$ whereas $x(2) = \begin{bmatrix} c \\ -c \end{bmatrix}$, c an arbitrary constant.

4.2 A Variation of Parameters Formula.

An important technique in studying the qualitative properties of solutions of linear and nonlinear ordinary differential and integral equations under perturbations is through the use of variation of parameters formula. In this section we develop such a formula for measure differential equations and consider some of its applications. The theorem below gives an analytic expression for solutions y of the nonlinear system (4.2) in terms of the solution x of the linear system (4.3) and the strength of the impulses a_k.

Theorem 4.3. Let the assumptions of Theorem 4.1 hold. If $x(t)$, given by (4.7) is the solution of (4.3), then for $t \in [t_{k-1} t_k)$, any solution $y(t)$ of (4.2) is given by

$$y(t) = x(t) + \int_{t_0}^{t} X(t,s)f(s,y(s))ds + \int_{t_0}^{t} X(t,s)g(s,y(s))du(s)$$

$$+ X(t,t_{k-1}) \sum_{i=1}^{k-1} a_i \left[\prod_{j=1}^{k-i} B_{k-j}^{-1} X(t_{k-j},t_{k-j-1}) B_i^{-1} A(t_i)(I_i+J_i) \right] \tag{4.14}$$

where

$$I_i = \int_{t_0}^{t_i} X(t_i,s)f(s,y(s))ds \; ; \; J_i = \int_{t_0}^{t_i} X(t_i,s)g(s,y(s))du(s) \; ,$$
$$i = 1,2,\ldots,k-1.$$

Proof. For $t \in [t_0, t_1)$, we have

$$y(t) = X(t,t_0)x_0 + \int_{t_0}^{t} X(t,s)f(s,y(s))ds + \int_{t_0}^{t} X(t,s)g(s,y(s))du(s). \tag{4.15}$$

As in the proof of Theorem 4.1,

$$y(t_1) = y(t_1-h) + \int_{t_1-h}^{t_1} A(s)y(s)du(s) + \int_{t_1-h}^{t_1} f(s,y(s))ds$$
$$+ \int_{t_1-h}^{t_1} g(s,y(s))du(s) \; ,$$

where $h > 0$ is sufficiently small. Letting $h \to 0_+$, we obtain from (4.15) and the fact that

$$\lim_{h \to 0_+} \int_{t_1-h}^{t_1} f(s,y(s))ds = 0,$$

$$y(t_1) = X(t_1,t_0)x_0 + \int_{t_0}^{t_1} X(t_1,s)f(s,y(s))ds$$

$$+ \int_{t_0}^{t_1-} X(t_1,s)g(s,y(s))du(s) + a_1\, A(t_1)y(t_1)+a_1(g(t_1,y(t_1))),$$

which, since B_1 is invertible yields

$$y(t_1) = B_1^{-1}[X(t_1,t_0)x_0 + I_1 + \int_{t_0}^{t_1-} X(t_1,s)g(s,y(s))du(s)$$

$$+ a_1(g(t_1,y(t_1)))] \qquad (4.16)$$

But, $B_1^{-1}\, X(t_1,t_0)x_0 = x(t_1)$ by (4.7). Further, since $x(t_1,t_1) = E$, we have

$$a_1 g(t_1,y(t_1)) = a_1 X(t_1,t_1)g(t_1,y(t_1)) = \int_{\{t_1\}} X(t_1,s)g(s,y(s))du(s).$$

Therefore,

$$y(t_1) = x(t_1) + B_1^{-1}\,(I_1 + J_1)$$

$$= x(t_1) + I_1 + J_1 + a_1\, B_1^{-1}\, A(t_1)(I_1+J_1). \qquad (4.17)$$

For $t \in [t_1,t_2)$, we know that

$$y(t) = X(t,t_1)y(t_1) + \int_{t_1}^{t} X(t,s)f(s,y(s))ds + \int_{t_1}^{t} X(t,s)g(s,y(s))du(s),$$

where $y(t_1)$ is determined by (4.17). Therefore,

$$y(t) = X(t,t_1)[x(t_1) + I_1 + J_1 + a_1\, B_1^{-1}\, A(t_1)(I_1+J_1)]$$

$$+ \int_{t_1}^{t} X(t,s)f(s,y(s))ds + \int_{t_1}^{t} X(t,s)g(s,y(s))du(s).$$

Now, for $t \in [t_1,t_2)$, $X(t,t_1)x(t_1) = x(t)$ by (4.7) and

$$X(t,t_1)(I_1+J_1) = \int_{t_0}^{t} X(t,s)f(s,y(s))ds + \int_{t_0}^{t} X(t,s)g(s,y(s))du(s)$$

by (4.5). Thus

$$y(t) = x(t) + \int_{t_o}^{t} X(t,s)f(s,y(s))ds + \int_{t_o}^{t} X(t,s)g(s,y(s))du(s)$$

$$+ a_1 X(t,t_1)\, B_1^{-1} A(t_1)(I_1 + J_1),\ t \in [t_1,t_2).$$

As above, it can be shown that

$$y(t_2) = x(t_2) + I_2 + J_2 + a_2\, B_2^{-1} A(t_2)\,(I_2 + J_2)$$

$$+ a_1\, B_2^{-1} X(t_2,t_1)\, B_1^{-1} A(t_1)\,(I_1 + J_1).$$

In general, for $t \in [t_{k-1}, t_k)$, (4.14) follows by induction.

q. e. d.

<u>Remark 4.3.</u> If $a_k = 0$ for each k, Theorem 4.3 reduces to the classical variation of parameters formula. When $A(t) = A$, a constant matrix, (4.14) takes the form

$$y(t) = x(t) + e^{tA}\{ \int_{t_o}^{t} e^{-sA}\ f(s,y(s))ds + \int_{t_o}^{t} e^{-sA}\ g(s,y(s))du(s)$$

$$+ \sum_{i=1}^{k-1} a_i \left[\prod_{j=1}^{k-i} B_{k-j}^{-1} \right] A(I_i' + J_i')\}, \qquad (4.18)$$

where $x(t)$ is determined by (4.11) and

$$I_i' = \int_{t_o}^{t_i} e^{-sA} f(s,y(s))ds\ ;\quad J_i' = \int_{t_o}^{t_i} e^{-sA}\ g(s,y(s))du(s),$$

$$i = 1,2,\ldots, k-1.$$

We now consider two applications of Theorem 4.3. The first is the bounded input bounded output (generally known as B I B O) stability results for measure differential equations. To simplify the analysis, we assume in equations (4.2) and (4.3) that $A(t) = A$, a scalar, $f \equiv 0$ and $g(t,y) = g(t)$ independent of y. Let $t_o = 0$.

<u>Theorem 4.4.</u> Suppose that

(i) $A < 0$;

(ii) there exist positive constants $m > 1$ and M such that

$$m < |b_i| \leq M, \quad i = 1,2,\ldots$$

where $b_i = a_i A - 1$;

(iii) $\sum_{i=1}^{\infty} | a_i g(t_i) | < \infty$.

Then, if the input g is bounded, so is the output y.

Proof. Let $t \in J$ be arbitrary. There is an index $k \geq 1$ such that $t \in [t_{k-1}, t_k)$. By Theorem 4.3, any solution $y(t) = y(t,t_0,x_0)$ of

$$Dy = Ay\, Du + g(t)\, Du$$

is given by

$$y(t) = x(t) + e^{tA} [\int_0^t e^{-sA} g(s)du(s)$$

$$+ \sum_{i=1}^{k-1} a_i A \, (\prod_{j=i}^{k-1} (1- a_j A)^{-1} I_i'] \tag{4.19}$$

where $x(t) = \left[\prod_{i=1}^{k-1} (1 - a_i A)^{-1} \right] e^{tA} x_0$ is the solution of $Dx = Ax\, Du$ and $I_i' = \int_0^{t_i} e^{-sA} g(s)du(s)$, $i = 1,2,\ldots,k-1$. By (ii), it is clear that the product $\prod_{i=1}^{k-1} (1-a_i A)^{-1}$ is bounded in absolute value as $k \to \infty$. Therefore, from (i), $x(t)$ is bounded on J. Next, since g is bounded on J, it follows from (i) and (iii) that

$$| \int_0^t e^{(t-s)A} g(s)du(s) | \leq \int_0^t e^{(t-s)A} |g(s)| \, ds + \sum_{i=1}^{\infty} |a_i g(t_i)| < \infty .$$

(i) and (iii) also imply that there is a constant $L > 0$ such that $e^{tA} I_i'$ $(1 \leq i \leq k - 1)$ is bounded by L on J. Hence by (ii) we have

$$| e^{t_{k-1}A} \sum_{i=1}^{k-1} a_i A \, (\prod_{j=i}^{k-1} (1 - a_j A)^{-1} I_i' |$$

$$\leq L \sum_{i=1}^{k-1} |1 + b_i| \left(\prod_{j=i}^{k-1} |b_j|^{-1} \right)$$

$$\leq L \sum_{i=1}^{k-1} (1 + M)\, m^{i-k}$$

$$\leq L(1 + M)\,(m-1)^{-1}$$

Therefore, from (4.19) we conclude that y is bounded on J.

Example 4.2. Let $A = -2$ and $g(t) = (t+1)^{-2}$, $0 \leq t < \infty$. Choose $t_k = k$, $a_k = -3 + k^{-1}$ for $k = 1,2,\ldots$. Then condition (ii) of Theorem 4.4 is satisfied as $3 < |b_k| < 5$, for $k \geq 1$. Also, since $\sum_{k=1}^{\infty} |a_k\, g(t_k)| \leq \sum_{k=1}^{\infty} 4k^{-2} < \infty$, condition (iii) also holds.

Our second application of Theorem 4.3 is concerning the asymptotic stability of the null solution of (4.2). We need the following lemma.

Lemma 4.1. Let u be a scalar function of bounded variation on $[t_o,T]$ and let r and p be non-negative scalar functions such that r is integrable and p is du-integrable on $[t_o, T]$. Then for any positive constants c and K, the inequality,

$$r(t) \leq c + \int_{t_o}^{t} Kr(s)ds + \int_{t_o}^{t} p(s)dv_u(s), \; t \in [t_o,T] \qquad (4.20)$$

implies

$$r(t) \leq c\, e^{K(t-t_o)} + \int_{t_o}^{t} e^{K(t-s)} p(s)dv_u(s), t \in [\, t_o,T \,] \qquad (4.21)$$

Proof. From (4.20), we clearly have

$$r(t) \leq y(t)$$

where $y(t)$ is the maximal solution of the integral equation

$$y(t) = c + \int_{t_o}^{t} K\, y(s)ds + \int_{t_o}^{t} p(s)dv_u(s), \; t \in [t_o,T]. \qquad (4.22)$$

Therefore, it suffices to prove that any solution of (4.22) satisfies the inequality

$$y(t) \leq (c+\delta)e^{K(t-t_o)} + \int_{t_o}^{t} e^{K(t-s)}\, p(s)dv_u(s) = z(t), \; t \in [t_o,T] \tag{4.23}$$

for any $\delta > 0$. This however will obviously follow if we show that $z(t)$ is a solution of

$$w(t) = (c+\delta) + \int_{t_o}^{t} K\, w(s)ds + \int_{t_o}^{t} p(s)dv_u(s), \; t \in [t_o,T] \tag{4.24}$$

that is to say

$$(c+\delta)e^{K(t-t_o)} + \int_{t_o}^{t} e^{K(t-s)}\, p(s)dv_u(s) = (c+\delta) + \int_{t_o}^{t} K\{(c+\delta)e^{K(s-t_o)} + \int_{t_o}^{s} p(\theta)e^{K(s-\theta)}\, dv_u(\theta)\}ds + \int_{t_o}^{t} p(s)dv_u(s). \tag{4.25}$$

We prove that

$$\int_{t_o}^{t} K\{(c+\delta)e^{K(s-t_o)} + \int_{t_o}^{t} p(\theta)e^{K(s-\theta)}dv_u(\theta)\}\, ds = (c+\delta)e^{K(t-t_o)} - (c+\delta) + \int_{t_o}^{t} e^{K(t-s)}p(s)dv_u(s) - \int_{t_o}^{t} p(s)dv_u(s). \tag{4.26}$$

Denoting the quantity on the left hand side of (4.26) by I, we obtain

$$I = \int_{t_o}^{t} K(c+\delta)e^{K(s-t_o)}ds + \int_{t_o}^{t} K\, e^{Ks}\{\int_{t_o}^{s} p(\theta)e^{-K\theta}\, dv_u(\theta)\}ds \tag{4.27}$$

$$= I' + I'' \text{ (say).}$$

Now

$$I' = (c+\delta)\, e^{K(t-t_o)} - (c+\delta) \tag{4.28}$$

and

$$I'' = \int_{t_o}^{t} \{\int_{t_o}^{s} p(\theta)e^{-K\theta}\, dv_u(\theta)\}d(e^{Ks}),$$

which, on integrating by parts, yields

$$I'' = \int_{t_0}^{t} e^{K(t-s)}\, p(s)\, dv_u(s) - \int_{t_0}^{t} p(s) dv_u(s). \qquad (4.29)$$

From (4.27) - (4.29), we see that (4.26) is established, which in turn establishes (4.25) and therewith the lemma.

q. e. d.

Theorem 4.5. Suppose that

(i) in equation (4.2), $A(t) = A$ is a constant matrix with all its eigenvalues having negative real parts;

(ii) for every $\varepsilon > 0$, there exist $\delta(\varepsilon)$, $T(\varepsilon) > 0$ such that

$$|f(t,y)| \leq \varepsilon |y| ,$$

whenever $|y| \leq \delta(\varepsilon)$ and $t \geq T(\varepsilon)$;

(iii) g satisfies

$$|g(t,y)| \leq p(t) ,$$

for $t \geq t_0$ and $|y| \leq H$, $H > 0$, where $p(t)$ is a non-negative function on J such that

$$\int_{t_0}^{\infty} p(s)\, dv_u(s) < \infty ; \qquad (4.30)$$

(iv) there exist positive constants P and Q such that

$$\lim_{k \to \infty} \left(\prod_{i=1}^{k} |B_i^{-1}| \right) = P < \infty$$

and

$$\lim_{k \to \infty} \sum_{i=1}^{k} |a_i A| \left(\prod_{j=i}^{k} |B_j^{-1}| \right) = Q < \infty.$$

Then, there exists $T_0 \geq \iota$ and δ_0 such that for every $t_0 \geq T_0$ and x_0 satisfying $|x_0| < \delta_0$, any solution $y(t) = y(t,t_0,x_0)$ of (4.2) satisfies $|y(t)| \to 0$ as $t \to \infty$. In particular, if (4.2) possesses the null solution, then it is asymptotically stable.

Proof. Let $t \in J$ be arbitrary and let $k \geq 1$ be such that $t \in [t_{k-1}, t_k)$. There exist positive constants M and α such that

$$|e^{tA}| \leq M e^{-\alpha t} \quad \text{for all } t \geq 0.$$

Let $0 < \varepsilon < \min\{\alpha K^{-1}, H\}$ where $K = M(Q+1)$. By (ii), choose $T(\varepsilon)$ and $\delta(\varepsilon)$ so that $T(\varepsilon) > t_0$ and $\delta(\varepsilon) \leq \varepsilon$. Now (4.30) implies that

$$e^{-ct} \int_{t_0}^{t} e^{cs} p(s)dv_u(s) \to 0 \quad \text{as} \quad t \to \infty,$$

for every $c > 0$, Therefore, there exists $T_0 \geq T(\varepsilon)$ such that

$$\int_{t_0}^{t} e^{-(\alpha-K\varepsilon)(t-s)} p(s)dv_u(s) < \delta(\varepsilon)(2K)^{-1} \tag{4.31}$$

for $t \geq T_0$. Let $\delta_0 = \delta(\varepsilon)\,(2PM)^{-1}$. Consider any $t_0 \geq T_0$ and x_0 satisfying $|x_0| < \delta_0$. Then, as long as $|y(t)| < \delta(\varepsilon)$, we have from (4.11), (4.18) and the conditions of the theorem,

$$|y(t)| \leq PM\,\delta_0\, e^{-\alpha(t-t_0)} + \int_{t_0}^{t} K\,\varepsilon\, e^{-\alpha(t-s)}|y(s)|ds$$
$$+ \int_{t_0}^{t} K\, e^{-\alpha(t-s)} p(s)dv_u(s).$$

Set $r(t) = e^{\alpha t}\,|y(t)|$ and apply Lemma 4.1 to obtain

$$r(t) \leq PM\,\delta_0\, e^{\alpha t_0}\, e^{K\varepsilon(t-t_0)} + \int_{t_0}^{t} K\, e^{\alpha s}\, e^{K\varepsilon(t-s)}\, dv_u(s),$$

that is,

$$|y(t)| \leq PM\,\delta_0\, e^{-(\alpha-K\varepsilon)(t-t_0)} + K \int_{t_0}^{t} e^{-(\alpha-K\varepsilon)(t-s)} p(s)dv_u(s). \tag{4.32}$$

Therefore,

$$|y(t)| \leq PM\,\delta_0 + K \int_{t_0}^{t} e^{-(\alpha-K\varepsilon)(t-s)} p(s)dv_u(s)$$
$$\leq \delta(\varepsilon)/2 + \delta(\varepsilon)/2 = \delta(\varepsilon),$$

from (4.31). Thus, the inequality $|y(t)| < \delta(\varepsilon)$ holds on J, which means that (4.32) holds on J, and so $|y(t)| \to 0$ as $t \to \infty$.

q. e. d.

Condition (4.30) above may be replaced by a more general condition

$$\int_t^{t+1} p(s)\, dv_u(s) \to 0 \quad \text{as} \quad t \to \infty.$$

Example 4.3. Condition (iv) of Theorem 4.5 is satisfied on $J = [0, \infty)$ if we choose $A = -1$, a scalar; $t_k = k$ and $a_k = 2(k^3-1)^{-1}$ for $k = 2,3,\dots$. Indeed, we have

$$\prod_{k=2}^{\infty} |B_k^{-1}| = \prod_{k=2}^{\infty} (1 - (2k^3 + 1)^{-1}) = 2/3,$$

and

$$\lim_{k\to\infty} \sum_{i=2}^{k} |a_i A| \, (\prod_{j=i}^{k} |B_j^{-1}|) \le \sum_{k=2}^{\infty} 2k^{-2} < \infty.$$

4.3 A Difference Equation.

Discrete-time systems constitute an independent branch of present day Mathematics. The purpose of this section is to discuss briefly how the study of the measure differential equation (4.3) reduces to that of a certain difference equation when u is a jump function.

As before, let $J = [t_0, \infty)$, $t_0 \ge 0$. Denote by J_{t_0} the discrete set of points $\{t_k = t_0 + k;\ k = 1,2,\dots\}$. Let

$$u(t) = \sum_{k=1}^{\infty} a_k H_k(t), \quad t \in J \tag{4.33}$$

where a_k and $H_k(t)$ are as in (4.4). From (4.3) and Theorem 2.1, it follows that

$$x(t_k) = x_0 + \sum_{i=1}^{k} a_i A(t_i)x(t_i) \tag{4.34}$$

for $k = 1,2,\dots$. Thus, on J_{t_0}, (4.3) reduces to the difference equation

$$\nabla x(t_k) = a_k A(t_k)x(t_k), \quad x(t_0) = x_0 \tag{4.35}$$

where the operator ∇ is defined by

$$\nabla x(t_k) = x(t_k) - x(t_{k-1}),\ k = 1,2,\ldots .$$

Let B_k be as in Theorem 4.1. The following result for the difference equation (4.35) is analogous to Theorems 4.1 and 4.2.

Theorem 4.6. (a) Let B_k be non-singular for each $k = 1,2,\ldots$ Then for any $x_0 \in R^n$, the (unique) solution of (4.35) through (t_0,x_0) on J_{t_0} is given by

$$x(t_k) = \left(\prod_{i=1}^{k} B_{k-i+1}^{-1}\right) x_0,\quad t_k \in J_{t_0} \qquad (4.36)$$

(b) Let B_j be singular for some $j \geq 1$. Then

(i) $x(t_j)$ does not exist in general ;

(ii) $x(t_j)$ is arbitrarily determined if $x_0 = 0$.

Remark 4.4. When $a_k = 0$ for all k, (4.35) reduces to $x(t) = 0$, $x(t_0) = x_0$ whose solution is $x(t) = x_0$, $t \in J_{t_0}$. In this case, (4.35) is stable but not asymptotically stable in general. If $a_k \neq 0$, the stability or the asymptotic stability is determined by the behaviour of the infinite product $\prod_{k=1}^{\infty} |B_k^{-1}|$. Indeed, from (4.36), it is clear that the null solution of (4.35) is stable if $\prod_{i=1}^{k} |B_i^{-1}|$ is bounded as $k \to \infty$ and it is asymptotically stable if $\left(\prod_{i=1}^{k} |B_i^{-1}|\right)^{-1} \to \infty$ as $k \to \infty$.

Next result is a discrete version of Gronwall's inequality.

Theorem 4.7. Let x and f be non-negative, scalar functions defined on J_{t_0} and let a_k be a sequence of non-negative real numbers such that

$$a_k f(t_k) < 1,\quad k = 1,2,\ldots .$$

Then for any positive constant c, the inequality

$$x(t_k) \leq c + \sum_{i=1}^{k} a_i\, f(t_i) x(t_i),\ t_k \in J_{t_0} \qquad (4.37)$$

implies

$$x(t_k) \leq c\, P_k^{-1} , \; t_k \in J_{t_o} ,$$

where P_k is the solution of the equation

$$P_k = (1-a_k f(t_k))\, P_{k-1},\; P_0 = 1.$$

In addition, if $\sum_{k=1}^{\infty} a_k\, f(t_k) < \infty$, then

$$x(t_k) \leq c\, P^{-1}, \quad t_k \in J_{t_o} ,$$

where $P = \prod_{k=1}^{\infty} (1 - a_k\, f(t_k))$.

Proof. Write the inequality (4.37) in the equivalent integral form

$$x(t_k) \leq c + \int_{t_o}^{t_k} f(s)\, x(s) du(s),\; t_k \in J_{t_o}$$

where u is as in (4.33) and proceed as in Theorem 2.3.

q. e. d.

4.4 General Results.

Here, we consider u to be any right-continuous function $\in BV(J,R)$. The measure du may be decomposed as

$$du = du_a + du_s \quad ,$$

where du_a is absolutely continuous and du_s is singular with respect to the Lebesgue measure.

Consider equations (4.3) and

$$Dy = A(t)y\, Du + F(t,y)Du \qquad (4.38)$$

where $x, y \in R^n$, $A(t)$ is an n by n matrix defined on J;F is defined on $J \times R^n$ with values in R^n, u is a right-continuous function $\in BV(J,R)$ and $A(t)x$ and $F(t,x)$ are du-integrable on J.

For any compact subinterval I of J, let NBV(I) denote the space of functions $f \in BV(I)$ which are normalized by the requirement that (i) f is right-continuous at each interior point of I and (ii) $f(\tau+) = 0$, where τ is the left end point of I.

On J, (4.38) is equivalent to

$$y(t) = x_0 + \int_{t_0}^{t} A(s)y(s)du(s) + \int_{t_0}^{t} F(s,y(s))du(s). \qquad (4.39)$$

Consider the more general integral equation

$$z(t) = \psi(t) + \int_{t_0}^{t} A(s)z(s)du(s) \qquad (4.40)$$

where $\psi(t)$ is a given function in $BV(J)$.

Theorem 4.8. Suppose that

(i) $A(t)\,\psi(t)$ is du-integrable ;

(ii) $A(t)$ is absolutely du-integrable ;

(iii) $\int_I |A(t)|\, d|u|_s(t) < 1$ for every compact $I \subset J$.

Then $z(.)$ defined by

$$z(t) = \lim_{k \to \infty} \Big[\psi(t) + \int_{\tau}^{t} A(s_1)\psi(s_1)du(s_1) + \dots$$

$$\dots + \int_{\tau}^{t}\int_{\tau}^{s_1}\dots\int_{\tau}^{s_{k-1}} A(s_1)A(s_2)\dots A(s_k)\psi(s_k)du(s_k)\dots du(s_1)\Big] \qquad (4.41)$$

is a solution of (4.40) for $t, \tau \in J$ and $t \geq \tau$.

Proof. Let I be any compact subinterval of J with left end point τ and let NBV(I) be the space of normalized functions $f \in BV(I)$ with $f(\tau) = \psi(\tau)$. For $\alpha \geq 0$, consider the space $NBV_\alpha(I)$ of functions $f \in NBV(I)$ with the norm

$$\|f\|_\alpha = \sup_{t\in I} \{V(f,[\tau,t])\exp(-\alpha \int_{\tau}^{t} |A(s)|d|u|(s))\}.$$

Then

$$\| f\|_\alpha \leq \sup_{t \in I} \{V(f,[\tau,t])\} \sup_{t\in I} \{\exp(-\alpha \int_\tau^t |A(s)|d|u|(s))\}$$

and

$$\| f\|_\alpha \geq \sup_{t \in I} \{V(f,[\tau,t])\} \inf_{t\in I} \{\exp(-\alpha \int_\tau^t |A(s)|d|u|(s))\}.$$

By absolute integrability of $A(t)$, there exist real numbers m and M such that

$$m \leq \exp(-\alpha \int_\tau^t |A(s)|d|u|(s)) \leq M, \quad t \in I.$$

Therefore, the norms $\|\cdot\|_\alpha$ are all equivalent and hence $NBV_\alpha(I)$ is complete for $\alpha \geq 0$, since $NBV(I)$ is so. Let $T: NBV_\alpha(I) \to NBV_\alpha(I)$ be the mapping defined by

$$(Tx)(t) = \psi(t) + \int_\tau^t A(s)x(s)du(s).$$

If $x,y \in NBV_\alpha(I)$, then since

$$V(\int_\tau^t A(s)du(s),I) \leq \int_I |A(s)|d|u|(s)$$

we have

$$V(Tx - Ty,[t_0,t]) \leq V(x-y,[t_0,t]) \int_{t_0}^t |A(s)|d|u|(s).$$

Splitting $d|u|$ into $d|u|_a$ and $d|u|_s$ and by condition (iii) there exists a real number α sufficiently large such that

$$\int_I |A(t)|\, d|u|_s(t) \leq 1 - \frac{1}{\alpha - 1}\ .$$

We further use the relation

$$\exp(\alpha \int_{t_0}^t |A(r)|\, d|u|_a(r)) - 1$$
$$= \int_{t_0}^t \alpha|A(s)|\, \exp(\alpha \int_{t_0}^s |A(r)|d|u|(r))d|u|_a(s)\ .$$

Now it is easy to verify that

$$\|Tx- Ty\|_\alpha \leq (\frac{1}{\alpha} + 1 - \frac{1}{\alpha-1})\, \| x-y \|_\alpha\ .$$

It follows that T is a contraction for $\alpha > 1$. Therefore, T has a unique fixed point which is the limit in NBV(I) of the sequence of iterates

$$x_1(t) = \psi(t)$$

$$x_k(t) = \psi(t) + \int_{\tau}^{t} A(s)\, x_{k-1}(s)du(s), \quad k = 2,3,\ldots .$$

This proves (4.41) for $t \in I$ and hence for $t \in J$ as I is arbitrary.

q. e. d.

From Theorem 4.8, it is clear that the solution $x(t) = x(t\,;\tau, x_0)$ of (4.3) is given by

$$x(t) = E(u, [\,\tau, t\,])\, x_0 \tag{4.42}$$

where

$$E(u,[\,\tau,t\,]) = \lim_{k\to\infty}\Big[E + \int_{\tau}^{t} A(s_1)du(s_1) + \ldots$$

$$\ldots + \int_{\tau}^{t}\int_{\tau}^{s_1}\ldots\int_{\tau}^{s_{k-1}} A(s_1)A(s_2)\ldots A(s_k)du(s_k)\ldots du(s_1)\Big]$$

E being the identity n by n matrix.

The matrix $E(u,[\,\tau, t\,])$ is called the evolution matrix generated by $A(t)$.

Remark 4.5. When $u(t) = t$, $E(u,[\,\tau,t\,]) = X(t,\tau)$, the transition matrix associated with $A(t)$. If $A(t) = A$, a constant n by n matrix, $E(u, [\,\tau,t\,]) = e^{(t-\tau)A}$.

Example 4.4. Consider the scalar equation

$$Dx = a(t)x\, Du$$

with $u = u_a + u_s$, where u_a is absolutely continuous and

$$u_s(t) = \begin{cases} 0 & \text{if } 0 \le t < \beta \\ \gamma & \text{if } t \ge \beta . \end{cases}$$

Let $a(t)$ satisfy $a(\beta)\,\gamma < 1$. Clearly

$$E(u,[0,t]) = E(u_a,[0,t]), \quad \text{if} \quad t < \beta$$

and for $t \geq \beta$,

$$E(u,[\beta,t]) = \lim_{k \to \infty} \Big[1 + \int_\beta^t a(s_1)du(s_1) + \ldots$$

$$\ldots + \int_\beta^t \int_\beta^{s_1} \ldots \int_\beta^{s_{k-1}} a(s_1)\ldots a(s_k)du(s_k)\ldots du(s_1)\Big].$$

Now

$$\int_\beta^t a(s_1)du(s_1) = \int_\beta^t a(s_1)du_a(s_1) + \int_\beta^t a(s_1)du_s(s_1)$$

$$= \int_\beta^t a(s_1)du_a(s_1) + \gamma\, a(\beta) \;;$$

$$\int_\beta^t a(s_1) \int_\beta^{s_1} a(s_2)du(s_2)du(s_1) = \int_\beta^t a(s_1) \int_\beta^{s_1} a(s_2)du_a(s_2)du_a(s_1)$$

$$+ \gamma\, a(\beta) \int_\beta^t a(s_1)du_a(s_1) + \gamma^2 a^2(\beta).$$

Therefore,

$$E(u,[\beta,t]) = [1 + \gamma\, a(\beta) + \gamma^2 a^2(\beta) + \ldots]$$

$$+ [1 + \gamma\, a(\beta) + \ldots] \int_\beta^t a(s_1)du_a(s_1)$$

$$+ [1 + \gamma\, a(\beta) + \ldots] \int_\beta^t a(s_1) \int_\beta^{s_1} a(s_2)du_a(s_2)du_a(s_1) + \ldots$$

$$= (1 - \gamma\, a(\beta))^{-1} E(u_a,[\beta,t]).$$

The evolution matrix satisfies the following properties:

(i) $E(u,[t,t]) = E, \quad t \in J$;

(ii) $E(u,[t,s])\, E(u,[s,r]) = E(u,[t,r]), \quad t, s, r \in J.$

(iii) $E(u,[\tau,t])$ is a function of bounded variation in t on every compact subinterval I of J with left end point τ.

To verify (iii), we note that

$$E(u,[s,t]) = \sum_{k=0}^{\infty} M^k(u,[s,t])$$

where

$$M^0(u,[s,t]) = E, \quad \text{and}$$

$$M^k(u,[s,t]) = \int_s^t A(r)M^{k-1}(u,[s,r])du(r), \quad k = 1,2,\ldots .$$

Theorem 4.9. Under the conditions of Theorem 4.8, $E(u,[\tau,t])$ satisfies the integral equation

$$E(u,[\tau,t]) = E + \int_\tau^t A(s)\, E(u,[\tau,s])\, du(s).$$

Proof.

$$\begin{aligned}\int_\tau^t A(s)E(u,[\tau,S])du(s) &= \int_\tau^t A(s) \sum_{k=0}^{\infty} M^k(u,[\tau,s])du(s) \\ &= \sum_{k=0}^{\infty} \int_\tau^t A(s)M^k(u,[\tau,s])du(s) \\ &= \sum_{k=0}^{\infty} M^{k+1}(u,[\tau,t]) \\ &= \sum_{k=0}^{\infty} M^k(u,[\tau,t]) - E \\ &= E(u,[\tau,t]) - E.\end{aligned}$$

q. e. d.

The following result is a variation of parameters formula and is more general than Theorem 4.3.

Theorem 4.10. Under the conditions of Theorem 4.8, any solution $y(t) = y(t;\, t_o,\, x_o)$ of (4.38) is given by the variation of parameters formula

$$y(t) = x(t) + \int_{t_o}^t E(u,[s,t])F(s,y(s))du(s) , \qquad (4.43)$$

where $x(t) = x(t;\, t_o,\, x_o)$ is the solution of (4.3) given by (4.42).

<u>Proof.</u> Substituting $y(t)$ from (4.43) into the right hand side of (4.39) and using Theorem 4.9, we have

$$x_0 + \int_{t_0}^{t} A(s)y(s)du(s) + \int_{t_0}^{t} F(s,y(s))du(s)$$

$$= x_0 + \int_{t_0}^{t} A(s)\Big[E(u,[t_0,s])x_0 + \int_{t_0}^{s} E(u,[r,s])F(r,y(r))du(r)\Big]du(s)$$

$$+ \int_{t_0}^{t} F(s,y(s))du(s)$$

$$= x_0 + \{E(u,[t_0,t])-E\}\, x_0$$

$$+ \int_{t_0}^{t} \{ \int_{r}^{t} A(s)E(u,[r,s])du(s)\} F(r,y(r))du(r)$$

$$+ \int_{t_0}^{t} F(s,y(s))du(s)$$

$$= x(t) + \int_{t_0}^{t} \{E(u,[r,t])-E\}\, F(r,y(r))du(r) + \int_{t_0}^{t} F(s,y(s))du(s)$$

$$= x(t) + \int_{t_0}^{t} E(u,[s,t])\, F(s,y(s))du(s)$$

$$= y(t).$$

q.e.d.

4.5 Notes.

Theorem 4.1 and 4.2 are taken from Pandit [32], and Theorems 4.3, 4.5 and Lemma 4.1 from Pandit [31]; see also Pandit [30] for additional results. Incidentally, Lemma 4.1 holds true for any bounded variation function u. Theorem 4.4 is due to Raghavendra [37], which also contains other interesting results. Section 4.3 is the work of Pandit and Deo [34]; for more information about discrete - time systems, see Sugiyama [49] and the references therein. Section 4.4 constitutes the recent work of Salkar and Deo [44].

CHAPTER 5

LYAPUNOV'S SECOND METHOD

The second method of Lyapunov plays a vital role in studying the qualitative behaviour of solutions of ordinary differential equations. This method depends on estimating the derivative of a scalar function along the solutions of the given differential equation. In this chapter, we extend Lyapunov's second method and employ it to investigate stability properties of solution of

$$Dx = f(t,x) + g(t,x)Du \tag{5.1}$$

with respect to the ordinary differential system

$$x' = f(t,x). \tag{5.2}$$

Section 5.1 contains definitions and basic results. The stability of asymptotically self invariant (ASI)sets, which is more general than that of usual invariant sets, is defined. In section 5.2, results on the exponential asymptotic stability and the uniform asymptotic stability of ASI sets relative to (5.1) are proved and Section 5.3 contains results on uniform and uniform asymptotic stability of solutions of (5.1) in terms of two measures.

5.1 Definitions and Basic Results.

Consider (5.1) and (5.2) where $f,g : R_+ \times S \to R^n$, R_+ is the non-negative real line and S an open set in R^n and u is a right-continuous function $\in BV(R_+,R)$. Assume that the solutions of (5.1) exist and are unique for $t \geq t_0$, $t_0 \in R_+$. Suppose that $f_x(t,x)$ exists and is continuous on $R_+ \times S_\rho$, $S_\rho = \{x \in R^n : |x| < \rho \ , \ \rho > 0\}$. For $(t_0,x_0) \in R_+ \times S_\rho$, let $x(t; t_0,x_0)$ denote the solution of (5.2). Consider the variational system

$$z' = f_x(t,x(t; t_0,x_0))z, \quad z(t_0) = E.$$

Let $\Phi(t; t_0,x_0)$ denote the fundamental matrix solution of the variational system.

Define the following classes of functions :

$\mathcal{K} = \{ \psi \in C[R_+, R_+] : \psi(t)$ is strictly increasing in t and $\psi(0) = 0\}$. Here $C[A,B]$ denotes the class of continuous functions from A to B.

$\mathcal{L} = \{\phi \in C[R_+, R_+] : \phi(t)$ is strictly decreasing and $\phi(t) \to 0$ as $t \to \infty\}$.

$\mathcal{A} = \{a \in C[R_+ \times [0, \rho), R_+] : a(t,r)$ is decreasing in t for each r, increasing in r for each t and $\lim_{\substack{t \to \infty \\ r \to 0}} a(t,r) = 0\}$.

$\mathcal{B} = \{H \in C[R_+ \times R_+, R_+] : \lim_{t \to \infty} (\sup_{t_0 \geq \tau} H(t,t_0)) = 0$, for some $\tau > 0\}$.

For any $h \in C[R_+ \times S, R_+]$ and $\sigma > 0$, let

$$p(h, \sigma) = \{(t,x) \in R_+ \times S : h(t,x) < \sigma \}.$$

<u>Definition 5.1.</u> The set $x = 0$ is said to be ASI relative to (5.2) if for every solution $x(t; t_0, 0)$ of (5.2), there is a $\phi \subset \mathcal{L}$ such that

$$|x(t; t_0, 0)| \leq \phi(t_0), \quad t \geq t_0.$$

<u>Definition 5.2.</u> The ASI set $x = 0$ is said to be

(i) <u>uniformly stable in variation</u>, if for each α, $0 < \alpha \leq \rho$, there exists a constant $M(\alpha)$ such that

$$| \Phi(t; t_0, x_0) | \leq M(\alpha), \ t \geq t_0$$

whenever $|x_0| \leq \alpha$;

(ii) <u>uniformly asymptotically stable in variation,</u> if for each α, $0 < \alpha \leq \rho$, there exists a $\phi_\alpha \in \mathcal{L}$ such that whenever $|x_0| \leq \alpha$,

$$| \Phi(t; t_0, x_0) | \leq \phi_\alpha(t-t_0), \ t \geq t_0;$$

(iii) <u>uniformly stable,</u> if there exists $a \in \mathcal{A}$ such that

$$|x(t; t_0,x_0)| \leq a(t_0, |x_0|), \; t \geq t_0 ;$$

(iv) <u>uniformly asymptotically stable,</u> if there exists $a \in \mathcal{A}$, $\phi \in \mathcal{L}$ and $H \in \mathcal{B}$ such that

$$|x(t; t_0,x_0)| \leq a(t_0,|x_0|)\phi(t-t_0)+ H(t,t_0), \; t \geq t_0;$$

(v) <u>quasi-equi-asymptotically stable,</u> if

$$|x(t; t_0,x_0)| \leq [a(t_0,|x_0|)+\beta]\phi(t-t_0)+ H(t,t_0), t \geq t_0$$

where a, ϕ, H are as in (iv) and β is a positive constant;

(vi) <u>exponentially asymptotically stable,</u> if

$$|x(t; t_0,x_0)| \leq [a(t,|x_0|)+\beta]e^{-\delta(t-t_0)}+H(t,t_0), \; t \geq t_0$$

where $a \in \mathcal{A}$, $H \in \mathcal{B}$ and $\beta > 0$, $\delta > 0$.

<u>Remark 5.1.</u> If $f(t,o) \equiv 0$ so that $x \equiv 0$ is the unique solution of (5.2), then the ASI set reduces to the usual invariant set.

Let h_0, $h \in C[R_+ \times S, R_+]$ and assume that $\inf \{h_0(t,x) : (t,x) \in R_+ \times S\} = 0$.

<u>Definition 5.3.</u> The differential system (5.2) is

(i) (h_0, h) - <u>stable</u>, if given $\varepsilon > 0$ and $t_0 \in R_+$, there exists $\delta = \delta(t_0, \varepsilon) > 0$ such that $h_0(t_0,x_0) < \delta$ implies $h(t,x(t)) < \varepsilon$, $t \geq t_0$;

(ii) (h_0,h) - <u>uniformly stable</u>, if (i) holds with δ being independent of t_0 ;

(iii) (h_0, h) - <u>equiasymptotically stable</u>, if (i) holds and for any $\eta > 0$, $t_0 \in R_+$, there exist numbers $\hat{\delta} = \hat{\delta}(t_0) > 0$ and $T = T(t_0, \eta) > 0$ such that $h_0(t_0,x_0) < \hat{\delta}$ implies $h(t,x(t)) < \eta$, $t \geq t_0 + T$;

(iv) (h_0, h) - uniformly asymptotically stable, if (ii) holds and the numbers $\hat{\delta}$ and T in (iii) are independent of t_0 ;

(v) (h_0, h) - unstable , if it is not (h_0, h) - stable.

It is easy to see that the above definition reduces to the definition of

(1) the well known Lyapunov stability of the invariant set $\{0\}$, if $h_0(t, x) = h(t,x) = |x|$;

(2) The partial stability of the invariant set $\{0\}$ if $h_0(t,x) = |x|$ and $h(t,x) = |x|_s = (\sum_{i=1}^{s} x_i^2)^{1/2}$, $s < n$;

(3) the stability of a set $M \subset R^n$, if $h_0(t,x) = h(t,x) = d(x,M)$, the distance of x from the set M ;

(4) the stability of the conditionally invariant set B with respect to the set A if $h_0(t,x) = d(x,A)$ and $h(t,x) = d(x,B)$ where A and B are such that $A \subset B$ and solutions starting in A remain in B ;

(5) the stability of the asymptotically self invariant set $\{0\}$ if $h(t,x) = |x|$ and $h_0(t,x) = a(t,|x|)$, $a \in \mathcal{A}$ or $h_0(t,x) = |x| + \phi(t)$, $\phi \in \mathcal{L}$.

Thus, an extremely unified study of stability concepts of different types of invariant sets is possible by employing two measures h_0 and h.

The two theorems that follow are basic in the investigation of stability properties of (5.1).

Theorem 5.1. Suppose that the ASI set $x = 0$ relative to (5.2) is uniformly asymptotically stable in variation. Then there exists a Lyapunov function $V(t,x)$ with the following properties:

(i) $V(t,x)$ is defined and continuous on $R_+ \times S_\rho$;

(ii) $|x| \leq V(t,x) \leq a(t, |x|)$, $(t,x) \in R_+ \times S_\rho$, $a \in \mathcal{A}$;

(iii) $|V(t,x)-V(t,y)| \leq M |x-y|$, $(t,x),(t,y) \in R_+ \times S_\rho$ where $M = M(\rho)$ is a Lipschitz constant ;

(iv) $$D^+V_{(5.2)}(t,x) = \lim_{h \to 0_+} \sup \frac{1}{h} [V(t+h, x+h\, f(t,x)) - V(t,x)]$$
$$\leq -c\, V(t,x) + \phi(t), \; (t,x) \in R_+ \times S_\rho$$

where $c > 0$ and $\phi \in \mathcal{L}$.

Theorem 5.2. Suppose that

(i) $m(t) \geq 0$ is a right-continuous function on R_+, with isolated discontinuities at t_k, $k = 1,2,\ldots$, $t_k > t_0$, such that

$$|m(t_k) - m(t_k-)| \leq \lambda_k, \text{ where } \sum_{k=1}^{\infty} \lambda_k < \infty;$$

(ii) $\hat{g} : R_+ \times R_+ \to R$ satisfies Caratheodory type condition, $\hat{g}(t,v)$ is non-decreasing in v for each t and

$$Dm(t) \leq \hat{g}(t, m(t)), \; t \in [t_i, t_{i+1}), \; i = 0,1,2,\ldots$$

where $Dm(t)$ is any Dini derivative of $m(t)$.

Then, $m(t_0) \leq v_0$ implies that

$$m(t) \leq r(t; t_0, v_0 + \sum_{k=1}^{\infty} \lambda_k), \; t \geq t_0,$$

where $r(t; t_0, v_0)$ is the maximal solution of

$$v' = \hat{g}(t,v), \; v(t_0) = v_0$$

existing on $[t_0, \infty)$.

5.2 Stability of ASI Sets.

Suppose that the set $x = 0$ is ASI relative to the unperturbed system (5.2). In this section, we give sufficient conditions for exponential asymptotic stability and uniform asymptotic stability of the ASI set $x = 0$ with respect to the system (5.1). Assume the following hypotheses:

(H_1) $f(t,x)$ and $f_x(t,x)$ are defined and continuous on $R_+ \times S_\rho$ and $f_x(t,x)$ is bounded on $R_+ \times S_\rho$;

(H_2) the ASI set $x = 0$ relative to (5.2) is uniformly asymptotically stable in variation ;

(H_3) $g(t,x)$ is measurable in t for each x, continuous in x for each t and

$$|g(t,x)| \leq p(t)|x| \quad \text{on } R_+ \times S_\rho,$$

where $p(t)$ is a dv_u-integrable function ;

(H_4) u is of type $\mathcal{G}$;

(H_5) the discontinuities of u occur at isolated points $\{t_k\}$, $t_1 < t_2 < \ldots < t_k < \ldots$ and are such that

$$|u(t_k)-u(t_k-)| \leq a_k \exp(-c(t_k-t_o)), \quad c > 0 \tag{5.3}$$

where a_k are constants and

$$\sum_{k=1}^{\infty} a_k\, p(t_k) < \infty ; \tag{5.4}$$

(H_6) $\int_t^{t+1} w(s)ds \to 0$ as $t \to \infty$, where

$$w(t) = \lim_{h \to 0_+} \sup \frac{1}{h} \int_t^{t+h} p(s)dv_u(s)$$

is the upper right Dini derivative of the indefinite integral

$$\int^t p(s)dv_u(s) ;$$

(H_7) there is $r > 0$ such that for each α, $0 < \alpha < r$, there exists $\gamma_\alpha \geq 0$ and a measurable function $p_\alpha(t)$ defined on $[\gamma_\alpha, \infty)$ such that

$|g(t,x)| \leq p_\alpha(t)$ for all $\alpha \leq |x| < r$ and $t \geq \gamma_\alpha$ and

$$G_\alpha(t) = \int_t^{t+1} p_\alpha(s)dv_u(s) \to 0 \quad \text{as } t \to \infty.$$

Define $Q_\alpha(t) = \sup\{G_\alpha(s) : t - 1 \leq s < \infty\}$. Then $Q_\alpha(t) \downarrow 0$ as $t \to \infty$.

Theorem 5.3. Let (H_1), (H_2), (H_3), (H_5) and (H_6) hold. Then the ASI set $x = 0$ is exponentially asymptotically stable with respect to (5.1).

The proof depends upon the following lemma.

Lemma 5.1. Let (H_1), (H_2), (H_3) and (H_6) hold. Then

$$D^+m(t) = \limsup_{h \to 0_+} \frac{1}{h} [\, m(t+h) - m(t)\,]$$

$$\leq M \rho\, w(t) + D^+V_{(5.2)}(t,x) \tag{5.5}$$

where $m(t) = V(t, y(t;\, t_0,x_0))$, $y(t;\, t_0,x_0)$ being the solution of (5.1), $V(t,x)$ is the Lyapunov function obtained in Theorem 5.1 and $M = M(\rho)$ is the Lipschitz constant for $V(t,x)$.

Proof. Set $x = y(t;\, t_0,x_0)$ and note that $y(t+h;\, t_0,x_0) = y(t+h, t, x)$, $h > 0$ in view of the assumed uniqueness of solutions of (5.1). Denoting the solution of (5.2) also by $x(s,t,x)$, we have by Lipschitzian property of V,

$$D^+m(t) = \limsup_{h \to 0_+} \frac{1}{h} [V(t+h, y(t+h;\, t_0,x_0)) - V(t, y(t,t_0,x_0))]$$

$$= \limsup_{h \to 0_+} \frac{1}{h} [V(t+h, y(t+h; t,x)) - V(t,x)]$$

$$\leq \limsup_{h \to 0_+} \frac{1}{h} [V(t+h, y(t+h;\, t,x)) - V(t+h, x(t+h;\, t,x))]$$

$$+ \limsup_{h \to 0_+} \frac{1}{h} [V(t+h, x(t+h;\, t,x)) - V(t+h, x + h\, f(t,x))]$$

$$+ \limsup_{h \to 0_+} \frac{1}{h} [V(t+h, x+h\, f(t,x)) - V(t,x)]$$

$$\leq M \limsup_{h \to 0_+} [\, \frac{1}{h} \, |y(t+h;t,x) - x(t+h;t,x)|\,] + D^+V_{(5.2)}(t,x)$$

$$\leq M \limsup_{h \to 0_+} \frac{1}{h} \{ \int_t^{t+h} |f(s,y(s;t,x)) - f(s,x(s;t,x))| ds$$

$$+ \int_t^{t+h} |g(s,y(s;t,x))| dv_u(s)\} + D^+V_{(5.2)}(t,x).$$

Therefore (5.5) follows from (H_3) and the fact that

$$\lim_{h\to 0_+} \sup \frac{1}{h}\left[\int_t^{t+h} |f(s,y(s;t,x))-f(s,x(s;t,x))|ds\right]$$

$$\leq K \lim_{h\to 0_+} \sup\left[\sup_{t\leq s\leq t+h} (|y(s;t,x)- x(s,t,x)|)\right] = 0,$$

since f is Lipschitzian in x.

q. e. d.

<u>Proof of Theorem 5.3.</u> The inequality (5.5) and Theorem 5.1 yield

$$D^+m(t) \leq M \rho w(t) + D^+V_{(5.2)}(t,x)$$

$$\leq M\rho w(t) + \phi(t)- c\, m(t),\ c > 0 \qquad (5.6)$$

Clearly, $\int_t^{t+1} [M\rho w(s)+ \phi(s)]\, ds \to 0$ as $t \to \infty$. Since $y(t; t_0,x_0)$ is continuous on $[t_{k-1},t_k)$, $k = 1,2,\ldots$, the inequality (5.6) implies, for each $t \in [t_{k-1}, t_k)$,

$$V(t,y(t; t_0,x_0)) \leq V(t_{k-1},y(t_{k-1},t_0\, x_0))\exp(-c(t-t_{k-1}))+H(t,t_{k-1}), \qquad (5.7)$$

where $H(t,t_{k-1}) = \int_{t_{k-1}}^{t} \exp(-c(t-s))[M \rho w(s) + \phi(s)]ds.$

At the points t_k, we have from (5.3) and (5.7),

$$V(t_k,y(t_k; t_0,x_0)) \leq V(t_k,y(t_k-;t_0,x_0))+M\rho\, p(t_k)|u(t_k)-u(t_k-)|$$

$$\leq V(t_{k-1},y(t_{k-1}; t_0,x_0))\exp(-c(t_k-t_{k-1})).$$

Now,

$$V(t,y(t;t_0,x_0)) \leq V(t_0,x_0)\exp(-c(t-t_0))+H(t,t_0),\ t \in [t_0,t_1)\ ;$$

$$V(t,y(t;t_0,x_0)) \leq V(t_1,y(t_1; t_0,x_0))\exp(-c(t-t_1))+H(t,t_1),\ t \in [t_1,t_2).$$

Hence, for $t \in [t_0, t_2)$, we have

$$V(t,y(t;t_0,x_0)) \le \{[V(t_0,x_0) + M\rho\, p(t_1)a_1]\exp(-c(t_1-t_0))$$

$$+ H(t_1,t_0)\} \exp(-c(t-t_1)) + H(t,t_1)$$

$$= [M\rho\, p(t_1)a_1 + V(t_0,x_0)]\exp(-c(t-t_0))+H(t,t_0),$$

since, by the definition of $H(t,t_0)$, it is obvious that

$$H(t,t_0) \exp(-c(t-t_1)) + H(t,t_1) = H(t,t_0).$$

In general, for $t \ge t_0$, it follows that

$$V(t,y(t;\, t_0,x_0)) \le [V(t_0,x_0)+M\rho \sum_{k=1}^{\infty} a_k p(t_k)]\exp(-c(t-t_0))+H(t,t_0).$$

Using the upper and lower estimates for $V(t,x)$ from Theorem 5.1, we obtain

$$|y(t;\, t_0,x_0)| \le [a(t_0,|x_0|)+\beta] \exp(-c(t-t_0))+H(t,t_0),$$

where $\beta = M\rho \sum_{k=1}^{\infty} a_k\, p(t_k)$. The desired conclusion now follows from the following observations

(i) $$\lim_{t\to\infty}\left[\sup_{t_0\ge 1} H(t,t_0)\right] = 0\,;$$

(ii) $H(t,t_0) \le \psi(t_0)$, $\psi \in \mathcal{L}$;

(iii) β can be made small by choosing k sufficiently large.

Remark 5.2. In the above proof, if we do not take $t_0 \ge T_k$ (k sufficiently large to make β as small as we wish), we could still conclude that the solutions of (5.1) tend to zero as $t \to \infty$, without the ASI set $x = 0$ being uniformly stable, that is, we can conclude that the ASI set is quasi-equi-asymptotically stable (see Definition 5.2 (v)).

Next result deals with the uniform asymptotic stability of the ASI set $x = 0$ relative to (5.1), under fairly general conditions.

<u>Theorem 5.4.</u> Let (H_1), (H_2), (H_4) and (H_7) hold. Then the ASI set $x = 0$ is uniformly asymptotically stable relative to (5.1).

<u>Proof.</u> Since the ASI set $x = 0$ relative to (5.2) is uniformly asymptotically stable in variation, the conclusions of Theorem 5.1 hold. In what follows, a, M, ϕ and c are the same as those appearing in Theorem 5.1.

Let $0 < \varepsilon < r$ be given. Choose $\delta = \delta(\varepsilon)$, $0 < \delta < \varepsilon$, so that

$$a(1, \delta) < \varepsilon/3. \tag{5.8}$$

Clearly, $\tilde{M}(t) = \int_t^{t+1} \phi(s)ds \to 0$ as $t \to \infty$, since $\phi \in \mathcal{L}$. Define $R(t) = \sup\{\tilde{M}(s) : t-1 \le s < \infty\}$. Then $R(t) \downarrow 0$ as $t \to \infty$. Let $N_1 = N_1(\varepsilon) \ge \gamma_\delta + 1$ be such that

$$R(t) < \varepsilon/3, \text{ for } t \ge N_1. \tag{5.9}$$

Let $\beta = \beta(\varepsilon) > 0$ be any number such that $\beta < \delta$. Select $T_1 = T_1(\varepsilon) \ge N_1$ so that

$$3\{M\, Q_\beta(T_1) + R(T_1)\} < \min\{c\,\beta, \varepsilon\}. \tag{5.10}$$

Let $|x_0| < \delta$ and $t_0 \ge T_1$. Then, if $y(t) = y(t; t_0\, x_0)$ is solution of (5.1) through (t_0, x_0), we assert that

$$|y(t; t_0, x_0)| < \varepsilon \quad \text{for } t \in [t_0, \infty). \tag{5.11}$$

Suppose not. Let T_2 be the first point larger than t_0 with $|y(T_2)| \ge \varepsilon$ (such a T_2 exists by the right-continuity of $y(t)$). Since the solution of (5.1) is unique to the right of t_0, there is a $\delta_1 \ge 0$ such that $\inf\{|y(t)| : t \in [t_0, T_2]\} = \delta_1$. If $\delta_1 = 0$ for some $\tilde{t} \in [t_0, T_2]$, then by uniqueness, $y(t) \equiv 0$ for all $t \ge \tilde{t}$ and the proof is trivial in this case. Suppose $\delta_1 > 0$. We now have $\delta_1 \le |y(t)| < r$ between t_0 and T_2. Let $t_1, t_2, \ldots t_k$ be the discontinuities of u in $[t_0, T_2]$. Using Theorem 5.1 and arguments similar to those as in the proof of Lemma 5.1, we get, in place of (5.5), the following inequality

$$D^+V_{(5.1)}(t,y(t)) \le -c|y(t)| + \phi(t) + M|g(t,y(t))||u'(t)| ,$$

which, on integration for $t \in [t_{i-1}, t_i)$ yields

$$V(t,y(t)) \le V(t_{i-1},y(t_{i-1})) - c\,\delta_1(t-t_{i-1}) + \int_{[t_{i-1},t]} \phi(s)ds$$

$$+ M \int_{[t_{i-1},t]} p_{\delta_1}(s)|u'(s)|ds \qquad (5.12)$$

At t_i, $V(t,y)$ clearly satisfies

$$|V(t_i,y(t_i)-V(t_i,y(t_i-))| \le M\,|g(t_i,y(t_i))|\,|u(t_i)-u(t_i-)|. \qquad (5.13)$$

Since $V(t,y)$ is continuous in t for each fixed y, we have from (5.12) and (5.13),

$$V(t_i,y(t_i)) \le |V(t_i,y(t_i))-V(t_i,y(t_i-))| + \lim_{h\to 0_+} V(t_i-h,y(t_i-h))$$

$$\le V(t_{i-1},y(t_{i-1})) - c\,\delta_1(t-t_{i-1}) + \int_{[t_{i-1},t]} \phi(s)ds$$

$$+M \int_{[t_{i-1},t]} p_{\delta_1}(s)|u'(s)|ds + M\,p_{\delta_1}(t_i)|u(t_i)-u(t_i-)|. \qquad (5.14)$$

From (5.12) and (5.14), it is easy to see, for $t \ge t_0$, where $t_0 < t_1 < \dots < t_N = t$, that

$$V(t,y(t)) \le V(t_0,x_0) - c\,\delta_1(t-t_0) + \int_{t_0}^{t} \phi(s)ds$$

$$+M\left[\sum_{i=1}^{N-1} p_{\delta_1}(t_i)|u(t_i)-u(t_i-)| + \sum_{i=1}^{N} \int_{[t_{i-1},t_i)} p_{\delta_1}(s)|u'(s)|ds\right]$$

$$= V(t_0,x_0) - c\delta_1(t-t_0) + \int_{t_0}^{t}\phi(s)ds + M\int_{t_0}^{t} p_{\delta_1}(s)dv_u(s)$$

$$\le V(t_0,x_0) - c\delta_1(t-t_0) + \int_{t_0}^{t} \tilde{M}(s)ds + M\int_{t_0-1}^{t} G_{\delta_1}(s)dv_u(s)$$

$$\leq V(t_o,x_o) + [M\, Q_{\delta_1}(t_o) + R(t_o) - c\,\delta_1]\,(t-t_o)$$

$$+ M\, Q_{\delta_1}(t_o) + R(t_o). \qquad (5.15)$$

Thus, at $t = T_2$, we obtain from (5.15) and (5.10) with $\beta = \delta_1$,

$$\varepsilon \leq |y(T_2)|$$

$$\leq V(T_2, y(T_2))$$

$$\leq V(t_o,x_o) + [M\, Q_{\delta_1}(t_o)\, R(t_o) - c\,\delta_1]\,(T_2 - t_o)$$

$$+ M\, Q_{\delta_1}(t_o) + R(t_o)$$

$$\leq a(t_o,|x_o|) + [M\, Q_{\delta_1}(T_1) + R(T_1) - c\,\delta_1]\,(T_2 - t_o)$$

$$+ M\, Q_{\delta_1}(T_1) + R(T_1)$$

$$\leq a(1,\delta) + \varepsilon/3 + \varepsilon/3$$

$$< \varepsilon/3 + \varepsilon/3 + \varepsilon/3 = \varepsilon.$$

This contradiction establishes (5.11), so that the ASI set $x = 0$ relative to (5.1) is uniformly stable. For the rest of the proof, fix $r_1 < r$ and choose $\delta_o = \delta(r_1)$ and $T_o = T_1(r_1)$. Fix $t_o \geq T_o$ and $|x_o| < \delta_o$. Then (5.11) implies that

$$|y(t;t_o,x_o)| < r_1 \quad \text{on } [t_o, \infty).$$

Let $0 < \eta < r_1$. Choose $\delta = \delta(\eta)$, $0 < \delta < \eta$ so that $3a(1,\delta) < \eta$ and $T_1(\eta) \geq N_1(\eta) \geq \gamma_\delta + 1$ so large that $R(t) < \eta/3$ for $t \geq N_1(\eta)$ and

$$3\{M\, Q_\delta(T_1) + R(T_1)\} < \min\{c\delta, \eta\}.$$

Select

$$T = \frac{3a(1, r_1) + 2M\, Q_\delta(1) + 3R(1) + 2c\delta T_1(\eta)}{2\, c\delta},$$

Then T depends on η but not on t_o, x_o. We claim that

$$|y(\tilde{t}_1; t_o,x_o)| < \delta(\eta) \tag{5.16}$$

for some $\tilde{t}_1 \in [t_o+T_1, t_o+T]$. If this were not true, then $\delta \leq |y(t; t_o,x_o)| < r$, for all $t \in [t_o+T_1, t_o+T]$. Let $y_o = y(t_o+T_1; t_o,x_o)$. Then by (5.15),

$$\begin{aligned} 0 < \delta < |y(t_o+T)| &\leq V(t_o+T, y(t_o+T)) \\ &\leq V(t_o+T_1, y(t_o,T_1)) \\ &\qquad + M[Q_\delta(t_o+T_1) + R(t_o+T_1) - c\delta](T-T_1) \\ &\qquad + M\, Q_\delta(t_o+T_1) + R(t_o+T_1) \\ &\leq a(t_o+T_1, |y_o|) - \tfrac{2}{3} c\delta(T-T_1) + MQ_\delta(1) + R(1) \\ &\leq a(1,r_1) - \tfrac{2}{3} c\delta(T-T_1) + MQ_\delta(1) + R(1) \\ &= 0, \end{aligned}$$

which is a contradiction, and (5.16) is established. Now (5.11) implies

$$|y(t; \tilde{t}_1, y(\tilde{t}_1; t_o,x_o))| < \eta \quad \text{on } [\tilde{t}_1, \infty),$$

since $\tilde{t}_1 \geq t_o + T_1 \geq T_1$ and $|y(\tilde{t}_1; t_o, x_o)| < \delta$. Therefore, by uniqueness of solutions of (5.1), it follows that

$$|y(t; t_o,x_o)| < \eta \text{ for } t \geq t_o+T, \text{ so that } |y(t; t_o,x_o)| \to 0$$

as $t \to \infty$, as η is arbitrary. Since T depends only on η and δ depends only on ε, the ASI set $x = 0$ relative to (5.1) is uniformly asymptotically stable.

q. e. d.

The following example shows the generality of the hypothesis (H_7).

<u>Example 5.1.</u> Define the scalar function $g(t,x)$ for $0 < t < \infty$, $0 \leq |x| < r$ as follows:

$$g(t,x) = \begin{cases} 2 & , \text{ for } t = 2n+1, \text{ n a positive integer} \\ \dfrac{1}{t(tx+1)}, & \text{for other values of } t. \end{cases}$$

Let $\gamma_{\delta_1} = \delta_1^{-1} + 1$. Then $p_{\delta_1}(2n+1) = 2$ and $p_{\delta_1}(t) = \dfrac{1}{t(t\delta_1 - 1)}$, otherwise. Choose

$$u(t) = \begin{cases} t^2 & , \text{ for } 0 \leq t < 2 \\ t^2 + \sum_{k=1}^{n} 2k & , \text{ for } 2n \leq t < (2n+1),\ n = 1,2,\ldots . \end{cases}$$

Then $p_{\delta_1}(t) \not\to 0$ as $t \to \infty$ and

$$\int_{\gamma_{\delta_1}}^{2n} p_{\delta_1}(s)\, dv_u(s) = 2 \int_{\gamma_{\delta_1}}^{2n} \frac{ds}{s\delta_1 - 1} + \sum_{k=n_0/2}^{\infty} \frac{1}{2k\delta_1 - 1}$$

where n_0 is the smallest integer greater than γ_{δ_1}. Therefore,

$$\int_{\gamma_{\delta_1}}^{\infty} p_{\delta_1}(s)\, dv_u(s) = \infty .$$

However, for $2n \leq t < 2n + 1$, we have

$$\int_{2n}^{2n+1} p_{\delta_1}(s) dv_u(s) = 2\int_{2n}^{2n+1} \frac{ds}{s\delta_1 - 1} + \sum_{k=n_0/2}^{n} \frac{1}{2k\delta_1 - 1} - \sum_{k=n_0/2}^{n-1} \frac{1}{2k\delta_1 - 1}$$

$$\to 0 \text{ as } n \to \infty .$$

Thus (H_7) holds. On the other hand (5.3) and (5.4) fail to hold.

5.3 Stability in Terms of Two Measures.

The advantage of studying stability properties by employing two different measures is that the results so obtained are most suited for a wide range of applications because of their generality and flexibility. This section is devoted to the investigation of the (h_0,h)-stability of (5.1). To this end we need a Lyapunov function $V(t,x)$ satisfying the following properties :

(1°) $V \in C[R_+ \times P(h,\sigma),R_+]$ and $V(t,x)$ is locally Lipschitzian in x for a constant $\bar{L} > 0$;

(2°) V is h-positive definite, that is

$$V(t,x) \geq b(h(t,x)), \ (t,x) \in P(h,\sigma), b \in \mathcal{K}.$$

Assume the following hypotheses :

($\tilde{H}_1$) $f \in C[R_+ \times S, R^n]$ and is Lipschitzian in x for a constant $\tilde{M} > 0$;

($\tilde{H}_2$) $|h(t,x)-h(t,y)| \leq \tilde{L}\, |x-y|$, $(t,x), (t,y) \in R_+ \times S$ and $h \in C[R_+ \times S, R_+]$;

($\tilde{H}_3$) $g(t,x)$ is defined on $R_+ \times S$, measurable in t for each x, continuous in x for each t and such that for $(t,x) \in P(h,\sigma)$,

$$|g(t,x)| \leq G(t)\, \psi\, (h(t,x)),$$

where $\psi \in \mathcal{K}$ and $G(t)$ is dv_u-integrable function ;

($\tilde{H}_4$) $\limsup_{\tau \to 0_+} \frac{1}{\tau} \int_t^{t+\tau} G(s)dv_u(s) = w(t)$;

($\tilde{H}_5$) the discontinuities of u occur at isolated points $\{t_k\}$, $t_1 < t_2 < \ldots < t_k < \ldots$, and are such that

$$|u(t_k) - u(t_k-)| \leq a_k ,$$

where a_k are constants.

The two lemmas below are concerned with obtaining a differential inequality with respect to (5.1), in terms of the Lyapunov function V.

<u>Lemma 5.2.</u> Let ($\tilde{H}_1$), ($\tilde{H}_3$) and ($\tilde{H}_4$) hold and suppose that there exists a Lyapunov function $V(t,x)$ satisfying properties (1°), (2°) and the differential inequality

$$D^+V_{(5.2)}(t,x) \leq g_1(t, V(t,x)), \ (t,x) \in P(h,\sigma) \tag{5.17}$$

where $g_1 \in C[R_+ \times R_+, R]$. Then, the differential inequality

$$D^{+} m(t) \leq g_2(t, m(t)) \tag{5.18}$$

is valid for those t where $m(t) = V(t,y(t; t_0,x_0))$ is defined, $y(t; t_0,x_0)$ being the solution of (5.1) through (t_0,x_0) and

$$g_2(t,r) = g_1(t,r) + \bar{L}\, w(t)\, \Phi(r),\ \Phi \in \mathcal{K}.$$

Lemma 5.3. Let $(\tilde{H}_1)$, $(\tilde{H}_2)$, $(\tilde{H}_3)$ and $(\tilde{H}_4)$ hold. Suppose that there exists a Lyapunov function $V(t,x)$ satisfying properties (1^0), (2^0) and the differential inequality

$$D^{+}V_{(5.2)}(t,x) + \mathcal{C}(h(t,x)) \leq g_1(t,V(t,x)),(t,x) \in P(h,\sigma), \tag{5.19}$$

where $\mathcal{C} \in \mathcal{K}$ such that $\mathcal{C}'(r)$ exists, $\mathcal{C}' \in \mathcal{K}$ and $g_1 \in C[R_+ \times R_+, R]$, $g_1(t,r)$ is nondecreasing in r for each $t \in R_+$. Then (5.18) holds almost everywhere.

The following lemma provides with the estimate of the jumps of $V(t,y(t; t_0, x_0))$ at the points t_k of discontinuity.

Lemma 5.4. (A) Let $(\tilde{H}_1)$, $(\tilde{H}_3)$ and $(\tilde{H}_5)$ hold. Suppose that there exists a Lyapunov function $V(t,x)$ satisfying property (1^0). If $y(t; t_0,x_0)$ is a solution of (5.1), then at the points of discontinuity $\{t_k\}$,

$$|m(t_k) - m(t_k-)| \leq \lambda_k , \tag{5.20}$$

where $m(t) = V(t,y(t; t_0,x_0))$ and $\lambda_k = \bar{L}\, \psi(\sigma)G(t_k)\, \alpha_k$.

(B) If in addition to the assumptions in (A), $(\tilde{H}_2)$ also holds, then at the points $\{t_k\}$, (5.20) is valid where

$$m(t) = V(t,y(t; t_0,x_0)) + \int_{t_0}^{t} \mathcal{C}(h(s,y(s; t_0,x_0)))ds.$$

Definition 5.4. The measure h_0 is said to be uniformly finer than the measure h, if there exists a constant λ and a function $\phi \in \mathcal{K}$ such that $h_0(t,x) < \lambda$ implies $h(t,x) < \phi(h_0(t,x))$.

Definition 5.5. The Lyapunov function $V \in C[R_+ \times P(h,\sigma),R_+]$ is said to be h_0- decrescent , if there exists a $\sigma_0 > 0$ and a function $\alpha \in \mathcal{K}$ such that $h_o(t_o, x_o) < \sigma_o$ implies $V(t,x) \leq \alpha(h_o(t,x))$.

We now come to the main stability results for the system (5.1).

Theorem 5.5. (A_1) Let the hypotheses of Lemma 5.2 and Lemma 5.4 (A) hold.

(B_1) Assume further that (i) $g_1(t,r)$ is nondecreasing in r for each t and $g_1(t,0) = 0$; (ii) $\sum_{k=1}^{\infty} \lambda_k$ converges ; (iii) h_o is uniformly finer than h; (iv) $V(t,x)$ is h_o-decrescent in $P(h_o,\sigma_o)$, $\sigma_o \in (0,\lambda)$ being such that $\phi(\sigma_o) < \sigma$. Then, the uniform stability of the null solution of the scalar equation

$$v' = g_2(t,v), \quad v(t_o) = v_o \tag{5.21}$$

($g_2(t,v)$ being the function obtained in Lemma 5.2) implies the $(\tilde{h}_o,h)$-uniform stability of the system (5.1), where

$$\tilde{h}_o(t,x) = \alpha(h_o(t,x)) + \beta(t), \beta \in \mathcal{L} \quad \text{such that } \beta(t_j) = \sum_{k=j}^{\infty} \lambda_k.$$

Proof. Let $0 < \varepsilon < \sigma$ and $t_o \in R_+$ be given. Since the null solution of (5.21) is uniformly stable, given $b(\varepsilon) > 0$, there exists a $\delta = \delta(\varepsilon) > 0$ such that

$$v(t; t_o, r_o) < b(\varepsilon) \ , \ t \geq t_o, \tag{5.22}$$

whenever $v_o < \delta$. Consider the function

$$\tilde{h}_o(t,x) = \alpha(h_o(t,x)) + \beta(t) \tag{5.23}$$

and note that $\inf \{ \tilde{h}_o(t,x); (t,x) \in R_+ \times S\} = 0$ and $\tilde{h}_o$ is uniformly finer than h. We assert that with the δ obtained above, the system (5.1) is $(\tilde{h}_o,h)$-uniformly stable. Suppose not. Then there exists a solution $y(t) = y(t; t_o,x_o)$ of (5.1) and a $\tilde{t} > t_o$ such that

$$\tilde{h}_o(t_o, x_o) < \delta \tag{5.24a}$$

$$h(\tilde{t}, y(\tilde{t})) = \varepsilon, \tag{5.24b}$$

and $h(t,y(t)) < \varepsilon$, $t \in [t_0,\tilde{t})$. Hence, in $[t_0,\tilde{t}]$, $h(t,y(t)) \leq \varepsilon < \sigma$, and by Lemma 5.2, we have

$$D^+m(t) \leq g_2(t, m(t)),\ t \in [t_i,t_{i+1}),t_{i+1} \leq \tilde{t} ,$$

$i = 0,1,2,\ldots$, where $m(t) = V(t, y(t))$. Since $|m(t_k) - m(t_k-)| \leq \lambda_k$, by Lemma 5.4 (A), from Theorem 5.2, we obtain for $t \in [t_0, \tilde{t}]$,

$$V(t,y(t)) \leq r(t; t_0,v_0 + \sum_{k=1}^{\infty} \lambda_k)$$

$$\leq r(t; t_0, v_0 + \sum_{k=0}^{\infty} \lambda_k) = r(t; t_0,v_0 + \sigma (t_0)).$$

Choosing $v_o = a(h_0(t_0,x_0))$, we have

$$V(t,y(t)) \leq r(t; t_0, \tilde{h}_0(t_0,x_0)),\ t \in [t_0,\tilde{t}].$$

Hence, using (5.24b) and the h-positive definiteness of V, we have

$$b(\varepsilon) = b(h(\tilde{t},y(\tilde{t})) \leq V(\tilde{t}, y(\tilde{t})) \leq r(\tilde{t},t_0,\tilde{h}_0 (t_0,x_0)) ,$$

which, in view of (5.23), (5.24a), the choice of $v_o = a(h_0(t_0,x_0))$ and (5.22) results in a contradiction.

q. e. d.

<u>Theorem 5.6.</u> Let the assumption (B_1) of Theorem 5.5 hold. Assume further that the hypotheses of Lemma 5.3 and Lemma 5.4(B) are satisfied. Then the uniform stability of the null solution of (5.21) implies the $(\tilde{h}_0,h)$-uniform asymptotic stability of (5.1), where $\tilde{h}_0(t,x)$ is the same function as in Theorem 5.5.

<u>Proof.</u> Let $0 < \varepsilon < \sigma$ and $t_0 \in R_+$ be given. Since the null solution of (5.21) is uniformly stable, the $(\tilde{h}_0,h)$-uniform stability of (5.1) follows as in Theorem 5.5. It remains to show that for every $\varepsilon > 0$, there exist $\delta_0 > 0$ and $T = T(\varepsilon) > 0$ such that $h(t,y(t)) < \varepsilon$, $t \geq t_0 + T$ whenever $\tilde{h}_0(t_0,x_0) < \delta_0$.

Let $\delta_0 = \delta(\sigma)$ where δ is the positive number associated with $(\tilde{h}_0,h)$-uniform stability of (5.1). Choose $T = T(\varepsilon) = b(\sigma)/\mathscr{C}(\delta(\varepsilon))$. Suppose that $\tilde{h}_0(t_0,x_0) < \delta_0$ and $h(t,y(t)) \geq \delta(\varepsilon)$, $t \in [t_0,T)$. Then, by Lemma 5.4(B) and Theorem 5.2, we have

$$V(t,y(t))+\int_{t_0}^{t} \mathscr{C}(h(s,y(s)))ds \leq r(t;\, t_0,\, v_0+\sum_{k=1}^{\infty} \lambda_k\,), \tag{5.25}$$

$t \in [t_0,\, t_0+T]$. Since $\mathscr{C} \in \mathcal{K}$, $\mathscr{C}(h(s,y(s))) \geq \mathscr{C}(\delta(\varepsilon))$, $s \in [t_0,\, t]$ and

$$r(t;\, t_0,v_0 \sum_{k=1}^{\infty} \lambda_k\,) \leq r(t;\, t_0,\, v_0 +\sum_{k=0}^{\infty} \lambda_k)= r(t;t_0,\tilde{h}_0(t_0,x_0)).$$

With $v_0 = a(h_0(t_0,\, x_0))$, (5.25) yields

$$V(t,y(t)) \leq r(t;\, t_0,\, \tilde{h}_0(t_0,x_0))- \mathscr{C}(\delta(\varepsilon))(t-t_0),\ t \in [t_0,T]. \tag{5.26}$$

By the uniform stability of the null solution of (5.21), we have $r(t;\, t_0,\, \tilde{h}_0(t_0,x_0)) < b(\sigma)$, $t \geq t_0$ whenever $\tilde{h}_0(t_0,x_0) < \delta_0$. Hence, using the h-positiveness of V, the choice of T and (5.26), we obtain

$$\begin{aligned} 0 < b(\delta(\varepsilon)) &\leq b(h(t_0+T,\, y(t_0+T))) \\ &\leq V(t_0+T,y(t_0+T)) \\ &\leq b(\sigma) - \mathscr{C}(\delta(\varepsilon))\,T = 0. \end{aligned}$$

This contradiction shows that there exists a $\tilde{t}_1 \in [t_0,t_0+T]$ such that $h(\tilde{t}_1,\, y(\tilde{t}_1)) < \delta(\varepsilon)$. Now, in view of the $(\tilde{h}_0,h)$-uniform stability of (5.1), it is clear that $h(t,y(t)) < \varepsilon$, $t \geq t_0 + T$, whenever $\tilde{h}_0(t_0,\, x_0) < \delta_0$.

q. e. d.

5.4 Notes.

Theorems 5.1 and 5.6 are respectively due to Leela [21] and Lakshikantham, Mitchell and Mitchell [20]. Theorem 5.3 and Lemma 5.1 are taken from Leela [22], which also contains a result on the uniform stability of the ASI set x = 0 relative to (5.1). Theorem 5.4, which is a slight modification of the corresponding result in Raghavendra and M.R. Rao [38], is contained in V.S.H. Rao [41], for other results such as uniform and uniform ultimate boundedness of impulsively perturbed systems, see V.S.H. Rao [42] and M.R. Rao and V.S.H. Rao [40]. The proofs of Lemmas 5.2, 5.3 and 5.4 are given in Leela [23], from which Theorems 5.5 and 5.6 are also adopted.

BIBLIOGRAPHY

[1] E.A. BARBASHIN, On stability with respect to impulsive perturbations, Differential'nye Uravneniya, 2(1966), 863-871.

[2] A.BLAQUIERE, Differential games with piecewise continuous trajectories. Lecture Notes in Control and Information Sciences No.3, Springer Verlag, New York (1977) 34-69.

[3] F. BRAUER, Nonlinear differential equations with forcing terms, Proc. Amer. Math. Soc. 15(1964), 758-765.

[4] J.CASE, Impulsively controlled D.G. in the theory and application of differential games (J.D. Grote Ed.). D. Reidel Publishing Company (1975).

[5] P.C. DAS and R.R. SHARMA, On optimal controls for measure delay-differential equations, SIAM. J. Control 9(1971),43-61.

[6] P.C. DAS and R.R. SHARMA, Existence and stability of measure differential equations, Czech. Math. J. 22(97), 1972,145-158.

[7] S.G. DEO and V. RAGHAVENDRA, Integral inequalities of Gronwall-Bellman type, unpublished lecture notes (1976).

[8] N. DUNFORD and J. SCHWARTZ, Linear operators, Interscience, New York, 1958.

[9] B.A. FLEISHMAN and T.J. MAHAR, Boundary value problems for a nonlinear differential equation with discontinuous nonlinearities, Math. Balkanica, 3(1973), 98-108.

[10] V.K.GORBUNOV, Differential impulsive games, Izv. Akad. Nauk SSSR, Tech. Kibernet (1973), no.4, 80-84. Translation Engineering Cybernetics 11(1973), no.4 pp.614-621.

[11] L.M. GRAVES, The theory of functions of real variables, McGraw-Hill, New York, 1956.

[12] A. HALANY and D. WEXLER, Teoria calitativa a sistemelor cu impulsuri, (Russian), Editura Academiei Republicii Socialiste Romania, Bucuresti, 1968.

[13] I. HALPERIN, Introduction to the theory of distributions, University of Toronto,Press, Toronto, 1952.

[14] E. HEWITT and K. STROMBERG, Real and Abstract Analysis, Springer Verlag, New York (1965).

[15] ITO ,K., On Stochastic differential equations, Mem Amer. Math. Soc. 4 (1951).

[16] GROH JIRGEN , A nonlinear-Stieltjes integral equation and a Gronwall inequality in one dimension Illinois Jour. Math. 24(1980), 244-263.

[17] GROH JIRGEN , On optimal control for nonlinear equations with impulses. An existence theorem, Sektion Mathematik, Friedrich-Schiller-Universitat, Jena (1980), 1-12.

[18] M.A. KRASNOSELSKII, Two remarks on the method of successive approximations, Usephi Mat. Nauk, 10(1955), 123-127 (Russian).

[19] V. LAKSHMIKANTHAM and S. LEELA, Differential and integral inequalities, Vol.I, Academic Press, New York, 1969.

[20] V. LAKSHMIKANTHAM, A.R. MITCHELL and R.W. MITCHELL, Differential equations on closed subsets of a Banach space, Trans.Amer. Math.Soc., 220 (1976), 103-113.

[21] S. LEELA, Asymptotically self invariant sets and perturbed systems, Annali di Matematica pura ed applicata, No.4,Vol.XCII (1972), 85-93.

[22] S. LEELA, Stability of measure differential equations, Pacific Jour. Math., 55 (1974), 489-498.

[23] S. LEELA, Stability of differential systems with impulsive perturbations in terms of two measures, Nonlinear Analysis, Theory, Methods and Applications, Vol.1, No.6(1977), 667-677.

[24] J. LIGEZA, On generalized solutions of some differential nonlinear equations of order n, Ann. Pol.Math. 31 (1975), 115-120.

[25] J. LIGEZA, On generalized periodic solutions of linear differential equations of order n, Ann. Pol. Math. 33(1977), 209-218.

[26] J.L. MASSERA, Contributions to stability theory, Ann. Math. 64(1956), 182-206, Erratum Ann. Math. 68(1958), 202.

[27] P. DE MOTTONI and A. TESI, On distribution solutions for nonlinear differential equations : nontriviality conditions, J.Diff. Eqs., 24 (1977), 355-364.

[28] S.G. PANDIT, Some problems in differential and integral equations containing measures, Ph.D. Thesis, University of Bombay, Bombay (1978).

[29] S.G. PANDIT, On the stability of impulsively perturbed differential systems, Bull. Austral. Math. Soc. 17(1977),423-432.

[30] S.G. PANDIT, Systems described by differential equations containing impulses : existence and uniqueness, Rev. Roum. Math. Pure. Appl 26(1981), 879-887.

[31] S.G. PANDIT, Differential systems with impulsive perturbations, Pacific Jour. Math. 86(1980), 553-560.

[32] S.G. PANDIT On differential systems containing measures, Bollettino dell'Unione Matematica Italiana (5) 17-A (1980), 451-457.

[33] S.G. PANDIT and S.G. DEO, Stability and asymptotic equivalence of measure differential equations, Journal of Nonlinear Analysis, Theory, Methods, and Applications, 3(1979), 647-655.

[34] S.G. PANDIT and S.G. DEO, Discrete-time systems via measure differential equations (To appear).

[35] T. PAVLIDIS, Optimal control of pulse frequency modulation systems, IEEE Trans. AC-11 (1966), 676-684.

[36] T. PAVLIDIS, Stability of systems described by differential equations containing impulses, IEEE Trans.AC-12(1967), 43-45.

[37] V. RAGHAVENDRA, Variation of constants formula for linear equations containing measures, unpublished manuscript.

[38] V. RAGHAVENDRA, and M. RAMA MOHANA RAO, On stability of differential systems with respect to impulsive perturbations, J. Math. Anal. Appl. 48(1974), 515-526.

[39] V. RAGHAVENDRA, and M. RAMA MOHANA RAO, Differential systems with impulsive perturbations and an extension of Liapunov's method (to appear).

[40] M. RAMA MOHANA RAO and V. SREE HARI RAO, Stability of impulsively perturbed systems, Bull. Austral. Math. Soc. 16 (1977), 99-110.

[41] V. SREE HARI RAO, Asymptotically invariant sets and stability of measure differential equations, Nonlinear Analysis, Theory, Methods and Applications 2(1978), 483-489.

[42] V. SREE HARI RAO, On boundedness of impulsively perturbed systems, Bull Austral. Math.Soc. 18(1978), 237-242.

[43] V. SREE HARI RAO, Note on asymptotic equivalence for measure differential equations,Bull. Austral. Math. Soc.18(1978), 455-460.

[44] S.M. SALKAR and S.G. DEO, On differential systems containing impulses (communicated).

[45] W.W. SCHMAEDEKE, Optimal control theory for nonlinear vector differential equations containing measures, SIAM. J. Control, 3(1965), 231-280.

[46] F.W. STALLARD, Functions of bounded variations as solutions of differential systems, Proc. Amer. Math.Soc., 13(1962),366-373.

[47] A. STRAUSS and J.A. YORKE, Perturbation theorems for ordinary differential equations, J. Diff. Eqs. 3(1967), 15-30.

[48] C.A. STUART , Differential equations with discontinuous non-linearities, Arch. Rat. Mech. Anal. 63(1976), 59-75.

[49] S. SUGIYAMA, Stability and boundedness problems on difference and functional difference equations, Mem. School of Sci.Engr., Waseda Univ. 33(1969), 79-88.

[50] W.M. WONHAM, Random differential equations in control theory. (131-212) A review paper in the monograph edited by A.T. Bharucha Reid 'Probabilistic Methods in Applied Mathematics' Vol.2, Academic Press, New York (1970).

[51] K. YOSIDA, 'Functional Analysis', Narosa Publishing House, New Delhi, 1974.

[52] S.T. ZABALISCHIN, Stability of generalized processes, Differencial'nye Uravneniya, Vol.2, (1966), 872-881.

INDEX